surfboard wax

A HISTORY

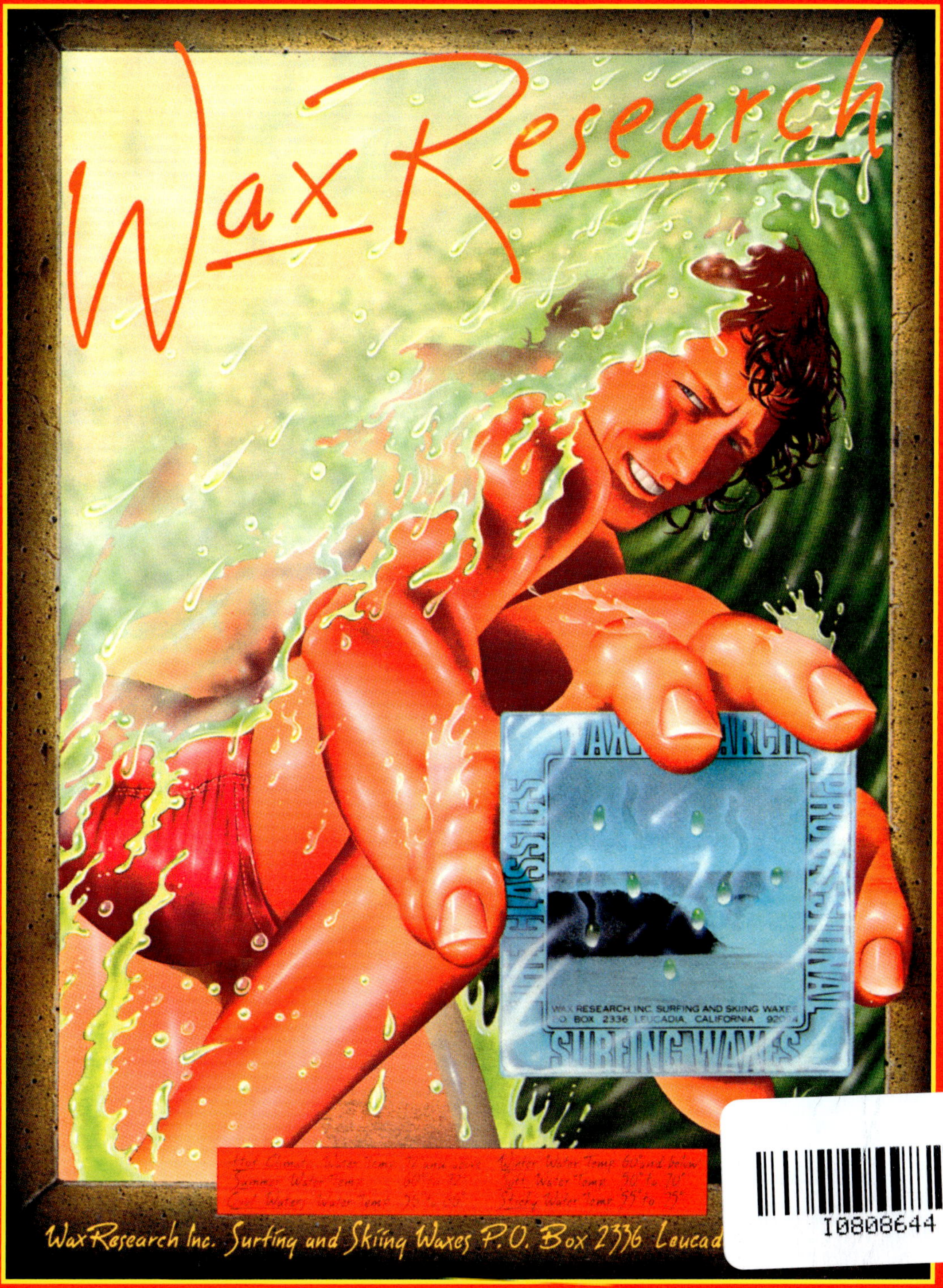

Jefferson "Zuma Jay" Wagner

Schiffer Publishing Ltd

4880 Lower Valley Road Atglen, Pennsylvania 19310

Preface

Surfboard wax—such a simple, basic item—is a tangible connection in the web that links us globally and through the generations. It binds us to the best days of our lives.

Social evolution and progressive social engineering quite often eliminate transition. Historical references are no longer necessary in daily life. Our sporting heritage is recorded, memorized, and brought forward only when reference is made to it.

Often, interpretations and opinions about history taint or fog the facts, unless recorded by pure, unabridged research work.

Advertising is our new medium, and the interpersonal communication forum is a keyboard. Rewards must be spontaneous. If some is good, then more is better. Profit motives take over and true sources fall to "progress."

The motivation to produce this book emerged from fond memories gained as an adolescent surfer. After forty-five years of experience in the ocean, I find that just as I fall asleep at night there is quite often a grin, because I was there in the beginning of a true California culture.

Jefferson Wagner
"Zuma Jay"
Malibu, California
2005

Acknowledgments

Nancy Schiffer, Editor, Schiffer Publishing
Larry Block Enterprises – Nelson
Tina Forde – computer editing
Felicia Martinez – photography
Ben Marcus – compiler, biographer, writer
Robert "Surfari Earth" Fisher – collector
Tsuyoshi Adachi and Ai Tamura – Japanese waxes
Eric Eason – East Coast waxes
Spencer Croul – Surfing Heritage Foundation
Skye and Rick Hodgson – Australian waxes
Candace Brown – wax list compiling
Tak Takamoto (CHP) – wax research details
Grant and Britt Galland
Sam Sciortino
Aqua East Surf Shop
Ava Wagner – daughter
Gail Marom – Middle East waxes
Bruce Grant – shaper
Kath Vicery – Australian waxes
Mark Huang – Communications Master
Elliot Wagner – undaunting patience

Other Schiffer Books on Related Subjects

Surf Art! Graphics and Memorabilia, by Rod Sumpter
Surfing, by Nancy N. Schiffer, in cooperation with San Francisco Airport Museums
Surf's Up: Collecting the Longboard Era, by Mark Blackburn

Library of Congress Control Number: 2006936628

Cover and book designed by: Bruce Waters
Type set in Geometric 231 Hv BT/Korinna BT

ISBN: 0-7643-2569-8
Printed in China

Published by Schiffer Publishing Ltd.
4880 Lower Valley Road
Atglen, PA 19310
Phone: (610) 593-1777; Fax: (610) 593-2002
E-mail: Info@schifferbooks.com

For the largest selection of fine reference books on this and related subjects, please visit our web site at
www.schifferbooks.com
We are always looking for people to write books on new and related subjects. If you have an idea for a book please contact us at the above address.

This book may be purchased from the publisher.
Include $3.95 for shipping.
Please try your bookstore first.
You may write for a free catalog.

In Europe, Schiffer books are distributed by
Bushwood Books
6 Marksbury Ave.
Kew Gardens
Surrey TW9 4JF England
Phone: 44 (0) 20 8392-8585; Fax: 44 (0) 20 8392-9876
E-mail: info@bushwoodbooks.co.uk
Website: www.bushwoodbooks.co.uk
Free postage in the U.K., Europe; air mail at cost.

Contents

Chapter One

In the Beginning - Parowax and Candles

Surfboard Traction From the Ancient Hawaiians to Mother's Cupboard

In the beginning, there was Parowax. Colorless and hopeless, Parowax was straight paraffin, boring stuff. It was white but it didn't smell good like coconut or lavender or anything else. It smelled like a petroleum product because that's what it was. Paraffin is the first byproduct in the refining process from crude oil to petroleum, and through most of the first half of the twentieth century, the oil companies scratched to find useful purposes for the stuff.

USEFUL – In the 1950s, homemakers kept a supply of Parowax on hand to seal canned items and to make candles. Their surfing youngsters discovered that the wax, while not perfect, provided traction on smooth surfboard surfaces. Although the oil industry was searching for new ways to market the petroleum product, surfboards were not mentioned on early labels. By the late 1960s, the label had evolved to include mention of the product's application to surfing.

THE MOTHER WAX, PAROWAX was a translucent bar of petroleum by-product that surfers liberated from the pantry and rubbed on their surfboards. *Photo by Dan Helms.*

Parowax was what mom used to seal canned fruit and to make candles. Parowax was not a surfing product and it was, according to Greg Noll, Dale Velzy, and a lot of others, "a bastard to get on your board." Applying Parowax to a surfboard was a ritual that required more than a little bit of the alchemist's skill, as it was a one-temperature wax, but no one ever figured out what that temperature was. Parowax went on hard and turned slick very, very fast. It didn't go on well or smell good like the more tempting modern versions, but at least young surfer kids of the 1940s, '50s and '60s weren't sent home from school sick after chewing and swallowing it.

THE BEGINNING – Skiers used Parowax on the bottoms of their sticks to go faster, while surfers used it to keep from sliding off the tops of theirs. The petroleum byproduct (note the Amoco logo), which could unstick a drawer or a zipper, cost about a quarter – this one, 26 cents.

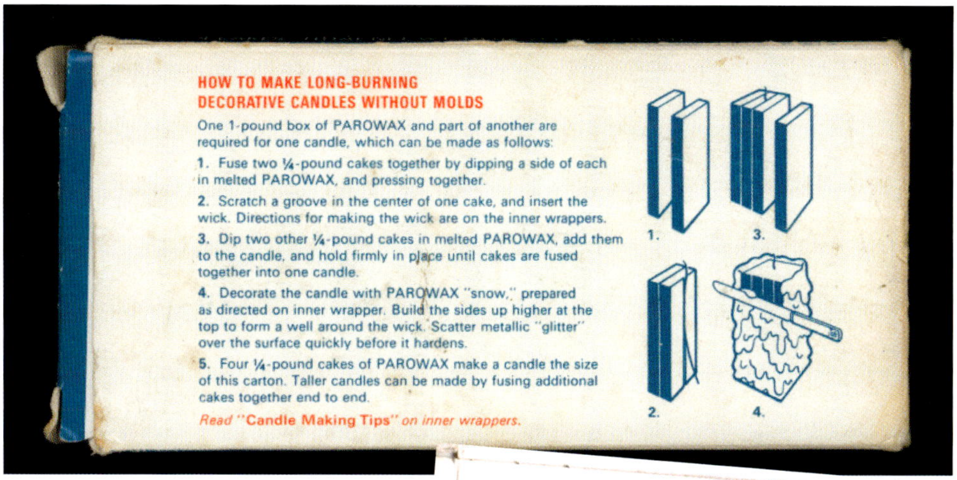

PROTOTYPE – A Parowax one-pound box label illustrates the proper way to make fancy candles, and inspires surfers. The 1/4 pound "cakes" were fused by melting them together. Perhaps mom accidentally dripped wax on sonny's surfboard and an industry was born. The oil companies finally picked up on the new niche market and by the late 1960s the label had evolved to include mention of the product's application to surfing.

ON THE SHELVES – Parowax with an updated label and Gulf Wax with a more contemporary look—both of which suggest the product's use on surfboards—are found on grocery shelves today in the canning sections.

HEY, MOM, GOT ANY LEFTOVER WAX? – Various vintage surfers claim to have been the first to try mom's sticky floor wax on their boards. The results were not entirely satisfactory. Floor wax was another petroleum byproduct.

Until the second half of the twentieth century, surfers were basically stuck with Parowax and other non-surfing products that came out of mom's cupboard: floor wax, candle wax, or canning wax. It was wax but it wasn't surf wax, and it all pretty much sucked from the time of the ancients until the '60s and early '70s when surfers like Mike Doyle and Rusty Miller and Frederick "Zog" Herzog began to put the "industry" in surf industry by making specially formulated surf wax. Until around 1966-67, waxing down your surfboard was something of an ordeal requiring time, elbow grease, and technique.

The Hawaiians. Pre-contact. Before Cook. What did they do for traction, you might ask yourself in an idle moment. Representing the Ancients, Greg Noll speaks from his own research when he states that "the early Hawaiians cut their boards with stone adzes, sanded them with coral and then sealed them either with coconut oil or pig slop. What did they use for traction? Hell if I know. Maybe the blood of captured slaves, maybe nothing. They probably just roughed up the decks with that coral or sand and kept them that way so there would be some friction and some traction."

Greg Noll's guess is as good as anyone's and it appears that the tradition of surfing on roughed-up decks continued well into the twentieth century. In the earliest decades of the twentieth century, surfers used sandpaper to rough up the varnish on their hardwood boards. These were the days before wetsuits or rash guards and one can only imagine how that felt on stomach, skin, hair, and a surfer's nipples.

A pad of corrugated automobile running board material provided a wide area of traction for this Pacific System Homes surfboard.

A well-worn Hawaiian board looks like a moonscape with intentional scratches and cuts covered by a sealer.

HAWAIIAN TRACTION – Ancient Hawaiians, up through the 1930s, gained traction by cutting their boards with stone adzes, sanding them with coral, and sealing them. The top board illustrates this "scratch" method. The Pacific System Homes surfboard shown at the bottom was decked with corrugated rubber automobile running board material, circa 1930. *Photo by Larry Hammerness, courtesy of Spencer Croul.*

In a statement that is hard to prove or disprove, a Palos Verdes surfer named Alfred Gallant Jr. claimed credit for the transition to wax after a trip across a sticky floor in 1935: "One night I came home from the Cove," Gallant recounted in a letter to *Longboard Magazine,* "and walked across the living room floor and my feet would almost stick on the wooden floors. My mother had waxed the floors with liquid wax. When I left to go to the Cove the next morning, I took the bottle of wax and put it on my board."

Gallant states he got in trouble for using his mom's expensive floor wax, and when she suggested he use the cheaper bars of paraffin that she used to can fruits and jellies, a revolution was hatched. The results were so frictive that Alfred and all his buddies immediately re-varnished their boards with smooth finishes and the paraffin revolution was ignited.

Hawaiian big-wave surfer Rickey Grigg reports similar experiences: "I used to use candle wax in the early '50s at Malibu. A couple of times back then I even used the wax my mom used to seal jars of jam. She never found out where they went. Some years later, surfers Mike Doyle and Rusty Miller invented the first surf wax in Encinitas, California, and advanced the sport of surfing by a giant leap akin to the one taken on the moon by Neil Armstrong several years later."

Waxing Poetic

Longboard,

In 1935 I started surfing at Palos Verdes Cove. In those days we made our own surfboards. We made them 11' or 12' long, of redwood or red cedar and varnished them spreading sand on the top when the varnish was still wet so that our feet wouldn't slip.

One night I came home from the Cove and walked across the living room floor and my feet would almost stick on the wooden floors. My mother had waxed the floors with liquid wax.

When I left to go to the Cove the next morning, I took the bottle of wax and put it on my board. When surfing, my feet just almost stuck on the board. It worked fine.

When I got home that night my mother was screaming at me. She said, "Where's the floor wax?" I told her we put it on the surfboards so our feet wouldn't slip on the board. Then mom said, "Use the stuff we use on the top of the jam jars." When my mom made the jam she would melt the wax and put it on top of the jam in the jars to preserve it. It acted like a cover so the jam wouldn't get sour. It was paraffin.

The next day I took the wax (paraffin) and put it on my board and all the other guys boards. That's how it started. We then removed the sand and re-varnished the boards so the sand would be gone and the wax was there.

In those days we were kneeling on the boards when paddling out. Our knees would get sore and sometimes bleed. With the wax on the board the bleeding didn't happen. My knees show little bumps where the bleeding took place.

My surfing buddies were among: John H. (Doc) Ball, Leroy Grannis, Fenton Scholes and Bud Morrissey. In those days there were only about fifty surfers in southern California. Our surfing places included Malibu, Santa Monica, Hermosa Beach, Redondo Beach, the entrance to Newport Harbor, San Onofre, and other spots. The boards in those days weighed about 125 pounds.

That was how waxing on the boards started.

Alfred E. Gallant,
Kaneohe, Oahu, HI

From all us of living in a slip-free world... thank you. —Ed.

MOTHER OF INVENTION – Alfred Gallant of Hawaii, who began surfing in 1935 at Palos Verdes Cove, turned a sticky, newly waxed kitchen floor into a "Eureka!" moment. His mom steered him toward the cheaper paraffin wax in the pantry.

Dawn

By Tom Morey with Ben Marcus

Somewhere back in the dim recesses of sometime, somebody recognized that a rough coating of wax on the deck of a surfboard provided a bit of extra traction—realized that in fact it worked quite well. But what about the different temperatures, amounts, application techniques? Who knew or cared, any more than thinking about polishing a train for greater speed?

Undoubtedly, the use of wax for traction came about as a result of using the stuff to seal dings. Wax is an elegant solution. No masking tape, messy varnish or resin, smelly catalyst, extensive drying time, or sanding itchy fiberglass. Just drip melted wax from a candle into a ding, maybe press it flat with the back of a cold spatula and presto! You're back in the water.

Then someplace along the line, cold wax was rubbed over a leaking break on a varnished deck, got felt and was liked! "Nifty! I think I'll try it rough like that on the whole deck!" We are all forever indebted to the guy who first acted on that thought. Just imagine being that first guy to wax up and paddle out!

By the time I entered the scene, 1952, everybody used wax. Parowax brand canning paraffin hadn't even become popular yet [as a surfing product].

That's because nobody even thought about BUYING wax. You got it free from your mother!! If she was a mom, she canned, made jellies and jams. And if she canned, she melted wax on top to seal in the goods.

Go for wax at a surf shop? There were no surf shops.

If your mom didn't have Parowax left over from canning, certainly she had a candle. And actually, candles were better because of their shape. ALL homes had emergency candles in the pantry because electricity was still not all that reliable.

CANDLE TECHNIQUE – Ava demonstrates how surfers in the '50s dripped hot wax from utility candles to wax their boards and fill in dings. Candles, says Tom Morey, were better than chunks of paraffin because of their shape. Many homes kept stocks of emergency candles "because electricity was still not all that reliable."

Sure, if quizzed, you might realize wax came from grocery stores. But what kind of a guy actually would consider walking into a store and asking for it? In the early '50s, that was mother's stuff. Buying wax would never even cross a guy's mind.

Normally you'd borrow a candle or hunk of wax from some other surfer—"borrow," of course, having nothing whatsoever to do with paying it back. Paying it forward, possibly...depending on who was asking. With borrowing came the obligation... the "in" responsibility of passing wax along to a good surfer or someone you liked, but pretending not to have any if some kook*—who'd then be out there in your way—asked.

**The word "kook" is a surfer's acronym for "Kicked Out Of Kaisers" (1970s), a Hawaiian surf spot; and later, in the '80s, for "Kid Out Of Kontrol."*

—Zuma Jay

Candle wax, Parowax...all the same. Who knew or cared one way or the other? Hey, if you were really desperate, you'd fish a used milk carton out of a trash can and rub your board with that. This was long before plastic-impregnated milk cartons and wax-impregnated paper cartons were common.

Or, you could heat up the milk carton over a beach fire, then rub it on. Beach fires were also far more common then, there being neither city or county restrictions against them, no fire pits to regulate where you could make one, no wet suits to keep you warm on the beach without one. Fire, a must!

The wax then (and still is) a drag. It melts into your car upholstery, flakes off, then melts to the roof, gets in your clothes, and on and on. Plus, seldom are you surfing on just wax. Invariably other surfers as a prank have tossed sand on your wax or kids and dogs have kicked it there by accident. A waxed deck is like a magnet. The hotter the waxed side of a board gets, the more sand it attracts.

Lay the board fresh from the water deck-down right in the sand and often you're better off. Whatever sand does affix to the deck washes right off. Or even better, prop up the middle, nose, and/or tail with kelp. Try to lean it up against something to stay totally sand-free and more often than not it gets knocked down—both dinged *and* covered with sand. If you don't surf you'd think this is a simple matter. It's not. When you finally come out of the water after waiting and waiting for just one more wave, you're usually beyond cold. The half-frozen brain is not thinking all that straight. Faulty choices are common.

"But doesn't a little sand here and there actually help the traction?"

I need to go stand by the fire.

—Tom Morey

The Mysterious First Product Actually Called Surf Wax

An Industry is Born

The traction market is all about marketing. If you happen to have access to a fairly complete set of *Surfer, International Surfing, Surf Guide, Australian Surfing Life* and *Surf Session* magazines going back to the beginning, it is possible to track the genesis of the traction market through the evolution of advertising.

Set the time machine for 1964 and open your issue of Surfer magazine, volume 5, no. 6, and you will find advertisements for a product ingeniously branded SURF WAX.

"Developed exclusively for surfboards," the ad proclaimed. The makers of SURF WAX "recognizing the need for a better surfboard coating, undertook to solve one of the oldest problems in the history of surfing. After two years of constant trials of various blends, we proudly offer you SURF WAX."

SURF WAX
developed exclusively for surfboards
Recognizing the need for a better surfboard coating, the Surf Wax Company undertook to solve one of the oldest problems in the history of surfing. After two years of constant trials of various blends, we proudly offer you **SURF WAX.**
SURF WAX is 'customized' for peak performance wherever used!
SURF WAX applies easily under the coldest and warmest conditions!
SURF WAX does not smooth over in cold water and does not flake or chip from your surfboard.
SURF WAX requires no hot-waxing, even when your board is new!
SURF WAX maintains a lasting adhesive quality which insures secure footing!
SURF WAX works best as its own base! Apply **SURF WAX** in a circular motion on a 'stripped' board for uniform coverage and personal preference.
SURF WAX is sold by leading surf shops throughout the United States!
DEALER INQUIRIES INVITED, CONTACT
SURF WAX
P.O. BOX 483
REDONDO BEACH, CALIFORNIA

WHO WAS IT? – A 1964 advertisement was Zen-like in its approach and product name, simply, SURF WAX. A company on the receiving end of P.O. Box 483 in Redondo Beach, California, announced the solution to "one of the oldest problems in the history of surfing"—how to keep from sliding off the board.

That "old problem" was how to keep from slipping off the board when the object was to stay attached as long, and in an upright position, as gracefully as possible. The snow skiing industry had developed waxing to an art, but with the goal of *reducing* friction on the bottoms of the boards rather than *increasing* friction on the tops. The medium, various forms of H2O, presented similar challenges based on temperature.

In the August 1965 issue of *Surfer* magazine there is an ad for "Customized SURF WAX" with new capabilities. The chemists at SURF WAX now recognized that different conditions needed different waxes, and they came up with three formulations: below 65 degrees, between 65 and 72 degrees, and between 72 and 80 degrees. Special orders were necessary for temperatures above 80 degrees and warm water SURF WAX could be applied on a "stripped board," which meant a board without a paraffin base.

SURF WAX offered all kinds of features and a Redondo Beach mailing address, but the origin of this wax—who made it, how they made it, and how much they sold—is a mystery wrapped in a paper wrapper. SURF WAX was first, but it would not last.

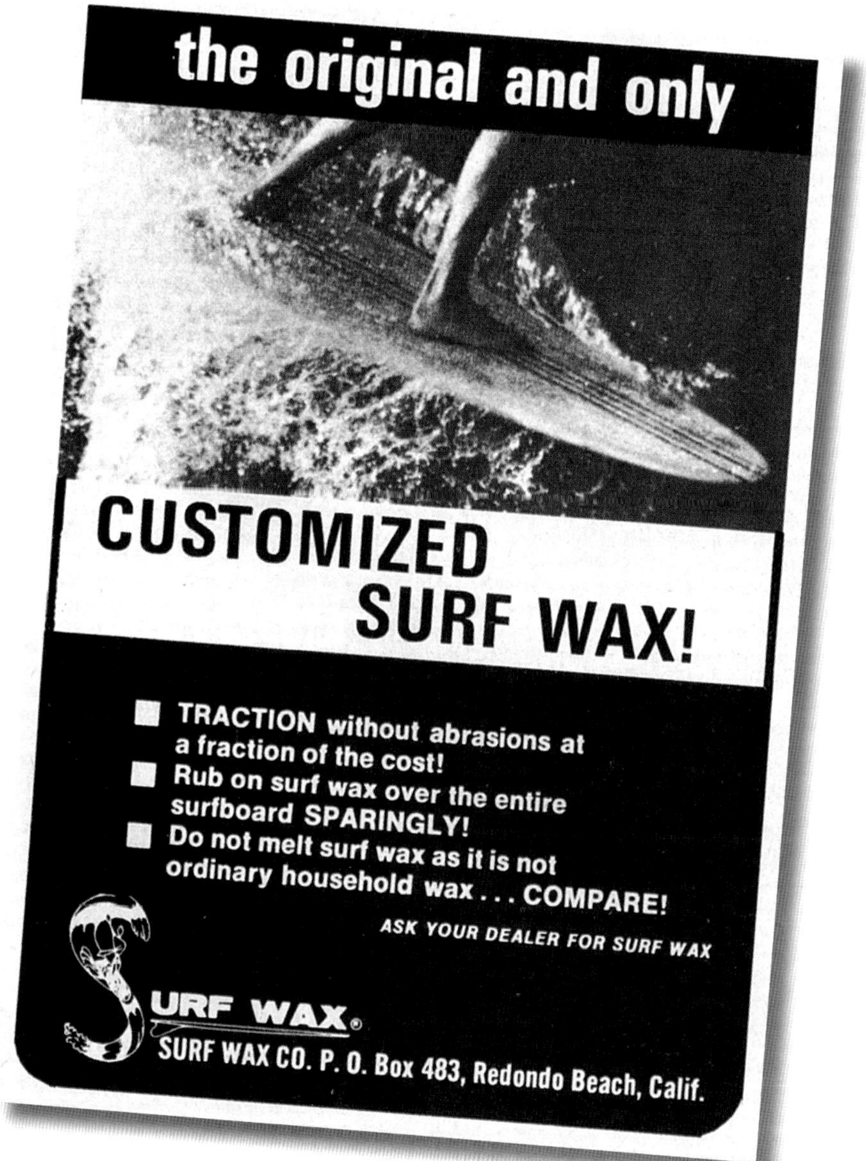

REFINEMENT – The Redondo Beach SURF WAX company bridges the gap between household wax and the nascent surf wax industry in this advertisement from *Surfer* magazine. "Do not melt surf wax as it is not ordinary household wax…" said the founders of "the original and only."

Chapter Two

The Surf-Industrial Complex - John Dahl and Surf Research

The simply named SURF WAX product was the opening shot in what has become a million-bar-a-month global industry. A number of companies joined the game through the years, only to retreat into oblivion like the Redondo Beach enterprise—pummeled by market forces, bad planning, or entrepreneurs who loved to surf more than they loved the drudgery of running a business.

The surfboard wax companies that fought their way to the top and survived are Surf Research (which evolved into Wax Research), Mr. Zog's, and Mrs. Palmers—referred to as "The Big Three." Two of the companies were born in California—Surf Research in Encinitas and Zog's in Goleta—while Mrs. Palmers began production in Australia, on Queensland's Gold Coast. All were started on shoestrings and built on ingenuity, doggedness, and an unshakable passion for the sport of surfing.

Surf Research was born in the middle 1960s and continues today in the form of Wax Research and Sticky Bumps. In forty years, Wax Research has transmogrified into one of the top three wax producers in the world. Today the company produces 20,000 bars a day from a 22,000 square-foot building in Carlsbad. The genesis of Surf Research into Sticky Bumps goes all the way back to 1966-67, when Mike Doyle, Rusty Miller, Garth Murphy, and Fred Ryan pooled their talents to form a surf accessory company.

COLOR CODED – Waxmate labels indicated the correct water temperature for which the wax was designed to be used. Blue was "cool" and green was "warm." John Dahl describes the early advertisements as "plain Jane," becoming more "silly and goofy but appealing."

FIRST ADVERTISEMENT– Founded in the mid-60s by Mike Doyle, Rusty Miller, Garth Murphy, and Fred Ryan, Surf Research was a surf accessory company that produced many things, but Waxmate was the product that stuck, so to speak. The first bars, said Steve Knorr, were made to "fit in a surfer's back pocket."

MEMORABLE MOTIF – The Surf Research ad campaign into the 1970s was innovative and successful, and often featured Mike Doyle, who captured perfectly the slightly mad world of surfing. It would not do to give an impression of being a button-down business. This ad appeared in *Surfer* issue 10, #6. *Image courtesy Wax Research and* Surfer *magazine.*

Mike Doyle described the early times in his book, *Morning Glass*: "After my frustrating experience with Catalina Swimwear, I was soured on the idea of working for another big company, but I started thinking I'd like to try my hand at running my own company. Two of my old buddies, Rusty Miller and Garth Murphy, had been thinking about doing the same thing, so back in Leucadia that spring, we decided to join forces and start a company together."

Take a look at early issues of *Surfer* magazine and Mike Doyle is all over the place: winning big surf contests, winning the 1964 *Surfer* poll, selling funny hats, ripping on a single ski, attaching his name to the Morey/Doyle softboards, surfing, modeling, selling, selling, selling. Doyle is one of the original hucksters who put the "industry" in "surf industry" and he was one of the sport's early marketing whizzes.

"I was the idea guy at Surf Research, the mad inventor," Doyle said. "Marketing and advertising. Anything we did worked. Our shop was right across the street from Hansen. He couldn't believe that you could surf all day and run a business."

THE FACE OF SURF RESEARCH – "Mike Doyle was the idea guy and helped to run the business and do promotions," said Rusty Miller, one of the founders of Surf Research. Both appear on this early ad featuring their chemical beaker logo.

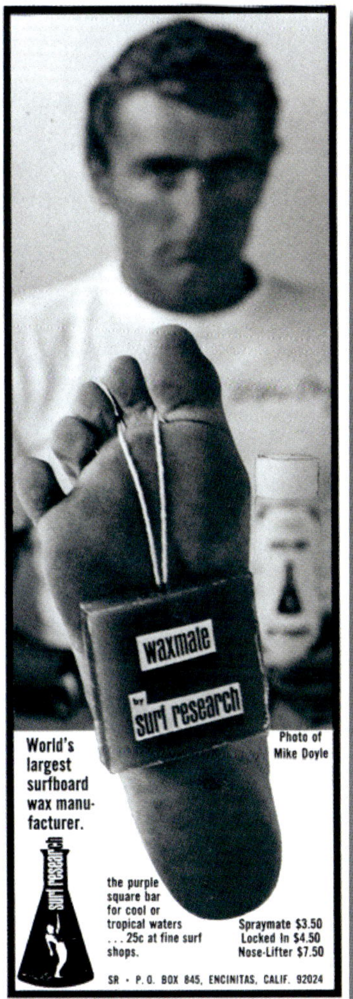

THE FOOT OF SURF RESEARCH – Waxmate makes a good toe tag for Mike Doyle. Steve Knorr revealed that one of the ingredients was motor oil. "It was brown and smelled funny, so we added a purple color."

PHARAOH RITUAL – What does it all mean? Why is Mike Doyle blowing smoke on the bar of wax? Whose t-shirt is it? How come it's so silly and looks so cool? It's a mystery. Surf Research stayed ahead of their more staid competitors in the ad department.

EXPANSION – Successful in its own backyard, Surf Research branched out to provide the surfing community in Europe with genuine California-made wax and other accessories. The company owners were not looking for mega-growth. "We didn't want to be surf moguls," Mike Doyle said. "We wanted to get our water time, be healthy and happy."

BI-COASTAL – Waxmate carved out a niche in the surf wax market and created brand name recognition. Surf Research also made waxes for other people with custom labels. "If there were five bars of wax on a shelf," John Dahl said, "four were out of our factory."

NEW NAME – The founders sold the name Surf Research and it faded away, yet the company lived on under the name Wax Research. Mike Doyle, here being attacked by kelp, sold his interest in 1969. Rusty Miller moved to Australia in protest against the Vietnam War.

BALMY AND PALMY – Wax Research Unlimited presented itself in a tropical mode, featuring six formulations, three of which sound like a summer salad—coconut, grape and banana. Astute businessman John Dahl kept the company sound.

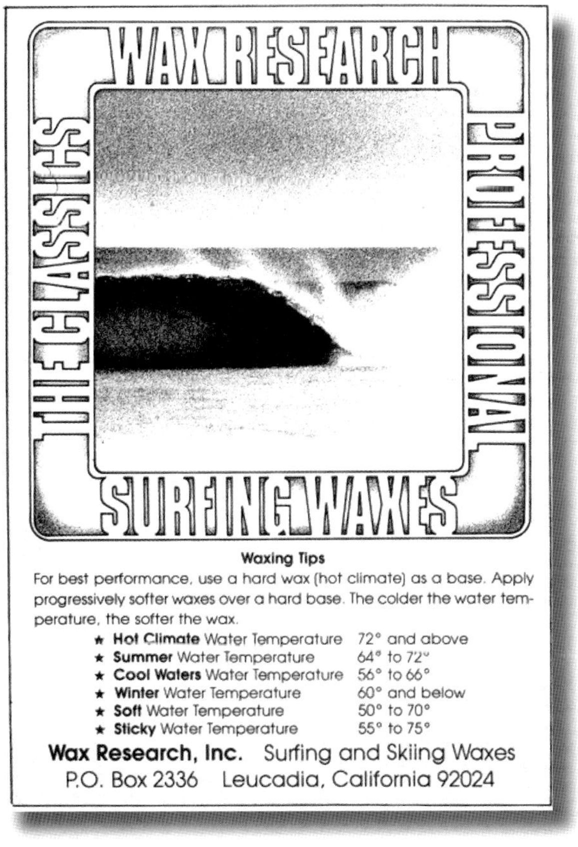

SURF AND SKI – Wax Research offered surf wax formulated for water temperatures high and low. "Apply progressively softer waxes over a hard base," the ad reads. "The colder the water temperature, the softer the wax."

Rusty Miller now lives in Australia. He was born in La Jolla and lived in Encinitas from the age of five. He has been surfing ever since he can remember, and he remembers buying paraffin at the gas station in the days before proper surf wax. Rusty Miller was one of Hansen's team riders along with Mike Doyle when the idea to start a surf accessory company was launched.

"There were four of us at the beginning: Fred Ryan, Mike Doyle, Garth Murphy and myself," Miller said. "We all had different roles. Mike Doyle was the idea guy and helped to run the business and do promotions. I was the salesman. I don't remember the very first sale, but I would go up and down the coast and then back, selling our wax and traction products to surf shops. Surf Research was a full-accessory company at first. We had the Barrecrafter line of surfboard racks from New York and we did surfing shorts. Doyle had worked for Catalina so we did some surf shorts using a sort of sailcloth. We had other products besides Waxmate. We had a product to compete with Slipcheck called Spraymate. But wax was the product that stuck and took off. We started with purple and then we had a cold water wax and a warm water wax. The first bars had a cellophane wrapper for a 3.2-ounce bar and on the back of every wrapper there was a fang and a fortune."

According to Garth Murphy, one of the favorite fortunes was, "Look at the moon, you may live there."

Their product line was extensive. "At Surf Research," Murphy said, "we made New Food, which was one of the first granolas. We sold surf racks, wetsuits, bikinis made by Mary Ryan, patch kits, Waxmate, Spraymate, and Lockmate, a cable lock for surfboards. The original Surf Research partners were Don Hansen, Mike Doyle, Rusty Miller and myself. We each pitched in $2,500 for a total of $10,000. I was financed in my 25% by Bill Engler, who made the Waxmate in a wax shed behind our offices, which is now the 'When In Rome' restaurant in Encinitas, about a block from Swami's (a southern California surf spot). We all worked for Hansen and he was a father figure for us. He wanted us to have our own company and the relative responsibilities. Our work ethic was so disgusting (because to be perfectly honest, it was very often fueled by mescaline, cherries and watermelon) that Hansen soon itched to sell his share. Fred Ryan and Ed Schumpert, a sculptor/shaper who did the Duke statue at the Huntington pier, had a company called Ocean Pacific making surfboards. Don Hansen traded Fred Ryan his 25% in Surf Research for the OP name. Don and Jim Jenks, Hansen's new manager, took OP and our swimwear ideas, some investment capital, and went on to make history and a great deal of money."

Knorr Candle Shop is a going concern in Del Mar. They have been in business since 1928, making candles and other things from pure beeswax. In the late '60s, Steve Knorr was in his early twenties and a surfer when he was approached by Mike Doyle and Rusty Miller, who had an idea to produce and distribute specially formulated wax for surfing tractions.

"They approached us first, I guess because we were in the candle business and they figured we knew about wax," Steve Knorr said. "But we deal with beeswax, and surf wax had always been paraffin and those are two different things. Paraffin is the first byproduct of refining crude oil and so it is fairly inexpensive and easy to get. I remember buying paraffin from Standard Oil or Gulf Richfield when I was learning to surf. But beeswax is the other extreme. It is a byproduct of making honey and it is expensive to produce. Back in the late '60s and early '70s, paraffin wasn't highly refined and we didn't have the sophisticated sources of paraffin they have now. The main thing Rusty and Mike wanted was wax that would go on easier and build bumps easier than the original paraffin. We tried mixing beeswax with paraffin but the mixture had too high a melting point so it was slicker on the board even than the Parowax. I don't know if this is a secret or not, but the original formulation for Waxmate was 30-weight Pennzoil motor oil mixed with paraffin. I don't know if you want to publish that. It was brown and smelled funny, so we added a purple color. They didn't want the combined petroleum products to smell like petroleum products, and all we had was grape. For the first several months or maybe a year or so it was purple with a grape scent."

Knorr Candle Shop produced the new wax at their factory for a couple of years.

"The process wasn't very glamorous," Steve Knorr said. "We would pour the hot mixture into these ice cube trays and let it harden and then break the bars out and wrap them. I guess the major idea was to make it so it would fit in a surfer's back pocket. Parowax was twice the size of a surfer's back pocket."

Parowax was twice the size but half the cost, and Steve Knorr was a little dubious about the business end of the surf wax business at first.

"Back then I couldn't imagine someone paying a quarter for a bar of surfboard wax," Knorr said. "You could buy a whole one-pound box of Parowax for a quarter. This was 1967 and a quarter was a lot of money for a bar of wax. But the market apparently was there because before too long Surf Research was placing some pretty big orders, from a few hundred bars a day to a few thousand. It kind of took off and it was fun to be a part of because it was surfing, so that was cool."

ORIGINAL FACTORY – The original partners produced the first Surf Research wax in this building in Encinitas across the street from board shaper and mentor Don Hansen. "Anything we did worked," Doyle said. "He [Hansen] couldn't believe that you could surf all day and run a business."

WRAPPING – A household iron was important equipment in the wax-wrapping process. The cellophane wrappers on each 3.2-ounce bar had a fortune. The favorite was "Look at the moon. You may live there."

According to John Dahl, Steve Knorr wasn't the only one who couldn't imagine paying a quarter for a bar of wax.

"On the West Coast we could hardly sell any at all—we were laughed out of the shops," Dahl said in a *Surfer* magazine article by Steve Barilotti. "Everybody was still using Parowax. Nobody wanted to pay a quarter for a three-ounce bar of surf wax when they could buy a whole pound of paraffin for 30 cents. It was only on the East Coast, where they were into all the accessories, that we made our sales. We shipped 75% of our wax back east."

In 1967, the first Surf Research ad appeared in Vol. 8, No. 3 of *Surfer* magazine. The ad says: "You don't need these…If you can ride the nose perfectly every time," and "these" referred to Spraymate, Nose Lifters, and Waxmate, "a purple square bar of wax that transfers to your board clear. Doesn't become mushy or slippery like others. At all fine surf shops. 25¢."

That first Surf Research ad was fairly plain Jane, but from there, driven by Mike Doyle's mind and presence, the Waxmate ads became increasingly clever, and are somewhere up there in surf marketing heaven, well below the Noll/Dora Da Cat campaign and "If You Don't Surf, Don't Start." But still up there. The Waxmate ads were simple and silly and goofy, but appealing, usually featuring Mike Doyle (sometimes wearing a veil of seaweed or pulling a crazy face). The ads were effective, and one of them was a little too effective.

"At Surf Research we used to make all our own advertisements for the surf magazines," Doyle wrote in *Morning Glass*. "They were simple and crude, but they were always fun and creative, and they usually brought a good response. For one of our ads in *Surfer*, we ran a picture of me with a t-shirt pulled over my head, like an Egyptian pharaoh's hood, blowing smoke over a bar of our surf wax. I don't even know what it was supposed to mean, some goofy, mystical notion we had.

"Not long after that ad appeared, Bill Engler was making wax bars in the backyard under a big mulberry tree, pouring the hot wax into molds. While he was working, mulberry leaves kept falling into the molds. Bill didn't let that slow him down, though—he had to get to Swami's while the surf was still glassy.

"Later that spring when I flew back to Florida on a promo tour for Hansen Surfboards, I took along some of our wax, thinking I would give out free samples to the surf shops. But apparently some redneck surf shop owner in South Florida had seen the ad of me blowing smoke over the wax, and when the bars arrived at his shop, he'd seen Bill's leaves embedded in them. He was sure we were putting marijuana leaves in the wax so kids could scrape it out and smoke it. He informed the police that I was smuggling marijuana, and on his tip I was stopped at the airport in Miami.

"The police were convinced I was from some California-Egyptian marijuana surf cult. When I finally figured out what they were talking about, I started laughing so hard I couldn't answer their questions. So one of the cops whipped out a bar of our purple surf wax, pointed to the leaves and said, 'What's this?'

"'It looks like mulberry leaves,' I said.

"The cop gave me a long, cynical stare, then said, 'Well, we're going to have it tested in our crime lab, and we'll let you know how it comes out. Now get outta here.'

"I never heard from them again."

The original SURF WAX faded from the magazines and while advertising was still dominated by spray-on traction ads for Control and El Gripo and Slipcheck, Waxmate began running regular and effective quarter-page ads, which began to blossom into full page ads. Within a couple of years they had a booming business, which the First Four took over completely.

"After a couple of years those guys took over their own production," Steve Knorr said. "I had split up from Mike and Rusty and gone out on my own. I knew the formula and was doing private label waxes for Surfboards Hawaii and Hobie. They had gone their own way with Waxmate."

That little house across the street from Hansen's served Surf Research well, but the arrival of John Dahl signaled changes that would require a change of space. Dahl, originally from La Jolla but now a longtime resident of North San Diego County, has surfed since 1954. Born John Baker, he took his wife's family name of Dahl.

"In 1967 I came to Surf Research and took the place of Garth Murphy, who moved to Australia," Dahl said. "I was the general manager, Fred Ryan became the owner, and Mike Doyle and Rusty Miller were still there as the surfers. Rusty was a salesman and Mike was the idea guy and I would say he was unbelievable idea guy. I believe Doyle came up with the first granola. We bought nuts and grains by the drum and packaged it as New Food and sold it to surf shops. Unfortunately, we were storing the New Food too close to the resin drums and after a while New Food had a resin flavor and that ended the product.

"Surf Research was a full surf accessory company at the time," Dahl said. "We were selling shorts and surf racks and nose lifters and a traction product called Slipmate along with Waxmate. I believe my contribution when I arrived at SR was the continual expansion for the surf wax part of the business. Surf Research was making five boxes of surf wax a day, pouring them in the backyard lawn area. When I came in we began to upgrade. We covered the entire backyard with corrugated fiberglass sheets and all of a sudden we were pouring twenty, thirty cases of wax a day. Pretty soon we were pouring so much wax we didn't have space to store all the raw materials, and that is when we moved to the old bank portion of the La Paloma Theater in Encinitas."

When Dahl came on board at Surf Research they also began experimenting with waxes for skiing, which was a much larger market at the time.

"Hilton Murphy, just back from two years in the Army—not Viet Nam but ski patrol in Alaska—came on board in 1969 as our experimental chemist," Dahl said. "He formulated many new waxes and tried a lot of different things. We changed all the formulas and developed the ski wax line, making a big leap. It wasn't a quantum leap. The quantum leap was Sticky Bumps in 1992."

Wax Research also made wax for other labels.

"We did a lot of OEM [original equipment manufacturer] waxes," Dahl said. "We did Laguna Wax and Frog House, Huntington Wax sold out at Jack's. We did Brotherhood out of Newport Beach for Paul Heussenstam. We also did Gordon and Smith's Orange Stuff. We created competition for ourselves but it was part of the growing volume of wax we were producing and we were the only people doing that at the time. This was pre-Zog or anyone else. We started the OEM stuff, which we still do. In contrast, Zog has never done that. There was a time where if there were five bars of surf wax on a shelf, four were out of our factory."

By 1968, there were a lot of traction products available to the general public, and not all of them were from Surf Research. The 1968 volume of *Surfer* magazine was loaded with competing ads for spray-on and rub-on traction products. There were full-page ads for Slipcheck and Control and El Gripo, with the Slipcheck and Control ads featuring doilies that surfers could apply to their boards for more groovy traction. The Surf Wax brand was hanging in there but their small, square ads were pretty square compared to what was coming out of Surf Research for Waxmate—the toe tag ad and the fish eye ad and the pharaoh ad and of course the infamous ad showing Fred van Dyke going left at Waimea, fin first.

The ad battle slacked off a bit in the 1969 volume of *Surfer*, but more than a half-dozen traction products were still available for those who were interested. Not everyone was. Greg Noll ended an era at Makaha on that giant day in December 1969, when he placed his mighty feet to the deck of a board that was waxed with good old-fashioned mama's paraffin.

"That first day at Waimea we used plain old paraffin canning wax and I'm pretty sure I was still using paraffin in 1969 at Makaha," Noll said. "Back then it was kind of an art form to get that stuff to work. Paraffin had one consistency so you put it on and if it got too hot you'd throw your board in the water to cool it down and if it got too cold you'd take some grains of sand and work it into the wax. Like I said, it was a bit of an art form but it worked well enough. Now they've got wax for cold water and hot water and Fiji and Hawaii and you can wax your ass and your nuts and all that bull. It's just a way for the wax makers to fill their coffers with pesos. We rode some big waves back then on paraffin, man."

But the times were about to be a-changin'. Surf Research was a fairly quick in-and-out for most of the founders. What had begun in 1967 was now doing well in the early '70s and some of the principals were ready to bolt to greener pastures.

"Toward the end of the '60s and into the early '70s," Rusty Miller said, "I was eager to move to Hawaii and Doyle wanted to get into real estate or something like that and we began talking about selling Surf Research. We were kind of like the Beatles, all going off in different directions. A friend named John Dahl was very anxious to buy the company and take over the management, and that was fine with me. John's dad and my dad were good mates and I'd known John all my life. He is an astute businessman and we all knew he would take the company we started and do good things with it. By that time Surf Research had a good pace. It wasn't a big wham bam success, but it had a nice even growth curve with a solid product. Mike and I got out in 1969. And it was a good time to sell."

Mike Doyle had joined up with Surf Research because he was disenchanted with the insincere corporate world he had gotten mixed up in at Catalina Sportswear. Like Woody Brown before him and many others, Doyle knew the dangers of getting sucked into the square world. Doyle described the same time with a different angle in *Morning Glass*: "Rusty, Garth and I knew we could easily double or triple the sales on almost every product we had at Surf Research. We could have sold twenty times as much purple wax. We couldn't keep up with the orders of our surf trunks, and that could have easily been expanded into a whole clothing line. But we'd watched Don Hansen go from a fun, fit, happy surfer living in his panel truck to…well, a businessman. We used Hansen as an example of what we didn't want to become. We didn't want to be surf moguls; we wanted to get our water time, be healthy and happy. Running a business had no appeal for us.

"To be fair to Don, though, he worked very hard and deserved the success he achieved. He would look at our business and wonder what in the world held the thing together. He used us an as example of what he didn't want his business to become.

"We hired a general manager for a while, John Baker [Dahl]. It was a pretty shrewd decision; he already had a wife and kids, so we figured he had no choice—he had to work.

"Meanwhile the whole country was going through some tough, gut-wrenching changes. Vietnam had turned into a bloody, full-scale war, and all of us had friends dying over there. A lot of us were becoming concerned about what we saw happening to the environment. People my age and younger were beginning to feel isolated from the older generation, and a mood of anger and despondency was setting in. We all had those feelings, but no one felt them stronger than Rusty Miller. Rusty understood history and politics better than the rest of us, and at times, talking about politics became an obsession with him. We would try to show Rusty that the world wasn't such a bad place, and he was just seeing it from an extreme perspective. But Rusty would resent that, thinking we were picking on him—which we were, because we loved him.

"Finally, Rusty said he was so disgusted with the United States, the only answer was to give up his U.S. citizenship and move to Australia. I tried to convince him there was no point in giving up his U.S. citizenship, that he could move to Australia and keep a dual citizenship, but he said he was so ashamed of his country, he wanted no part of it any more.

"That fall," Doyle said, "Rusty packed up his few belongings and headed for Australia. After he left, the synergy of our little group fell apart. Without Rusty, Surf Research just wasn't that much fun any more. Garth, Bill, and I talked things over. We all agreed it was time for us to dissolve the partnership. We remained close friends—in fact, Garth, Bill, and I became even closer—we just weren't business partners any longer. We sold the wax works to a guy in Colorado who wanted to make ski wax. He changed the name from Surf Research to Wax Research, and it was over. I had some money in my pocket, and it was time for a freedom run."

According to Garth Murphy, the Cream song "Pressed Rat" summed up the mood at Surf Research, particularly regarding the mogulizing of Don Hansen and Jim Jenks: "Check the lyrics: 'Pressed Rat and Warthog closed down their shop. They didn't want to, t'was all they had got,'" Murphy said in a communication from "deepest Baja."

"We broke up Surf Research because of the stress of the clothing business and twice-weekly trips to the LA garment district with hairball highway runs at 75 mph and no center divider. Rusty went to Kauai to live at Taylor Camp and Mike Doyle moved to Oregon to be a lumberjack and coldwaterman and single skier. I went to Australia, fell in love, and opened Surf Research Australia with Rusty Miller and Bill Engler, headquartered by Byron Bay. We did well in Australia for two or three years but had a horrible time distributing from Byron by train. There was no national truck service and our distributor in Sydney and Melbourne paid late and then did not pay at all. Wax is heavy for the price and supply was difficult, with raw materials coming from Shell in Indonesia. I sold the Surf Research name to an Aussie mate who let it dwindle and die, Waxmate formula, labels, and all. The smell of melting wax eventually washed from our skins and psyches, and we all lived happily ever after."

John Dahl was there to pick up the reins.

"What happened was Rusty and Mike sold out and Fred Ryan finally sold Surf Research to his brother-in-law in Denver," John Dahl said. "They moved the whole business there to focus on ski waxes. At the time we had a product called US Ski Wax and we did private label waxes for Head and AMS out of the Tahoe area. So Surf Research kind of spun off and went to Denver and left a bit of a vacuum. Waxmate faded and then disappeared. That is when Hilton Murphy and I started Wax Research. That was 1971-1972. Wax Research was totally focused on producing specialty waxes to begin with—carving wax and jewelry wax—but surf wax slowly rose to the top until that was all we were doing. Hilton was a chemist and could do all the unique formulas. We also were doing a lot of custom ski waxes at the time and we still do. But as the volume picked up, the focus turned to surf wax."

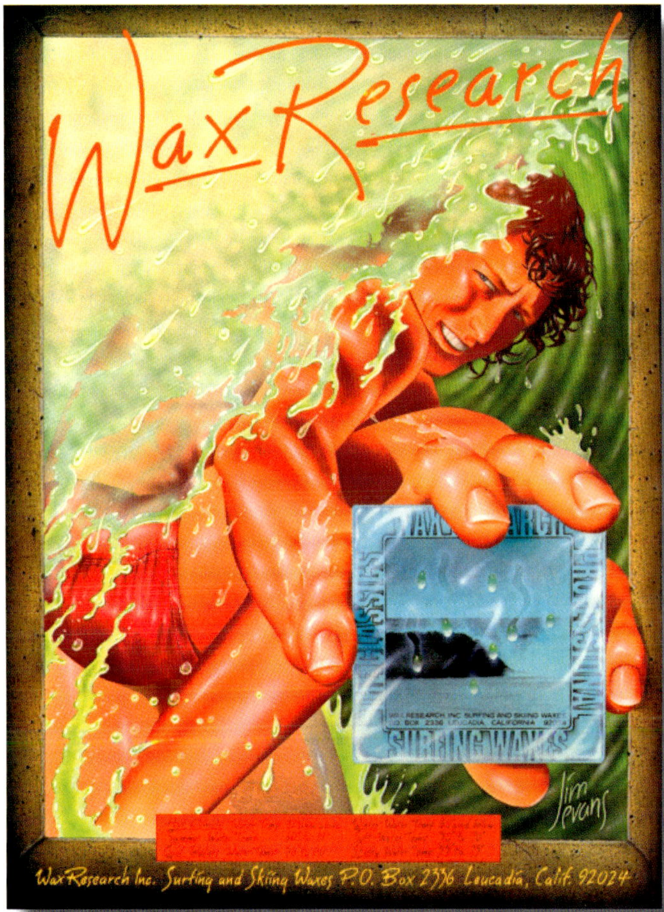

PROGRESSION – Wax Research ads became more sophisticated as the company grew. Mike Doyle's goofy "simple and crude"—but effective—photos gave way to colorful graphics that translated well onto stickers and t-shirts. This dynamic artwork is by Jim Evans.

Sticky Bumps Breakthrough

True technological advances in the surfboard wax industry are few and far between. One such breakthrough was John Dahl's creation of a product called Sticky Bumps.

"The biggest transition I can think of was 1992 when we formulated the new Sticky Bumps," Dahl said. "Frankly, the invention of Sticky Bumps was serendipitous. We came up with Sticky Bumps while we were experimenting with a variety of materials for an entirely unrelated product. It was kind of like a witch's brew: What happens when I mix this with that? Sticky Bumps came out by accident, but as soon as I saw it and used it, I knew there was nothing like it.

"Wax Research owns the trademark Sticky Bumps. There was a lawsuit over the brand, which we won in the six figures.

"Sticky Bumps actually started in 1973. The wax and the name went through its own metamorphosis. It was a Day-Glo style wax in the late '70s. The wax went from Sticky to Sticky Wax to New Sticky to Sticky Bumps. And then in 1992 the formulation of Sticky Bumps changed.

"The Sticky Bumps style of wax is thirty years old. I have all the labels that show the evolution of the brand."

Dahl's product, he agrees, is similar to another one made under the Mr. Zog's label, called Quick Humps. Dahl was first on the market.

"We are making similar waxes," Dahl said, of the Zog wax, "but nobody has figured out our production technique. That is what continues to set Sticky Bumps apart from other waxes. Not just ingredients, but how we put it together. That is the one big trade secret we have here. Process is number one. Formula is number two."

BUMPY WAX – Wax Research's successful product that eventually became Sticky Bumps was born back in 1973 as Sticky Wax. John Dahl changed the formula significantly in 1992, and he was first on the market with this uniquely textured surf wax.

Chapter Three

What's In a Name? Sex Wax Rises in the West

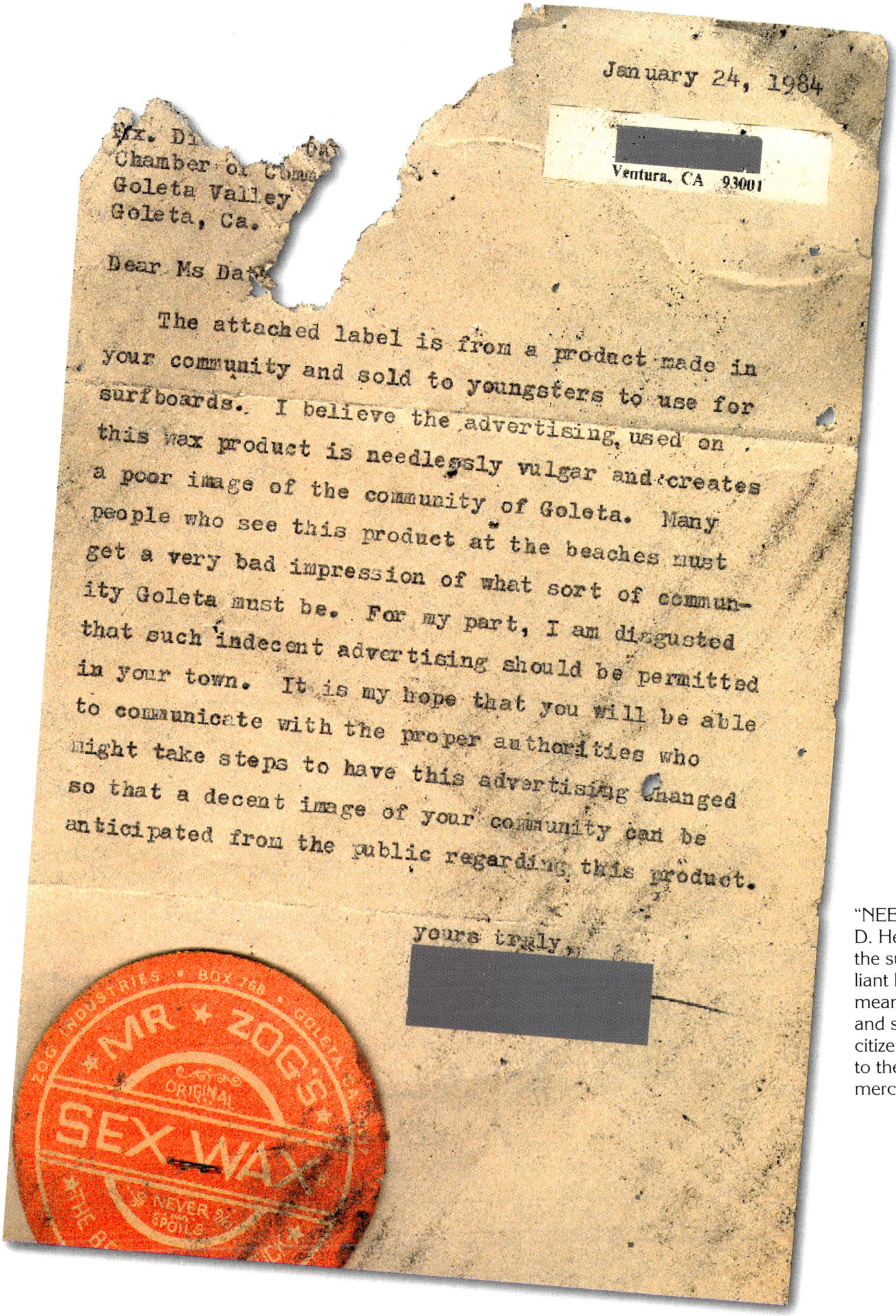

January 24, 1984

Ventura, CA 93001

Mx. Di
Chamber of Comm
Goleta Valley
Goleta, Ca.

Dear Ms Da

The attached label is from a product made in your community and sold to youngsters to use for surfboards. I believe the advertising used on this wax product is needlessly vulgar and creates a poor image of the community of Goleta. Many people who see this product at the beaches must get a very bad impression of what sort of community Goleta must be. For my part, I am disgusted that such indecent advertising should be permitted in your town. It is my hope that you will be able to communicate with the proper authorities who might take steps to have this advertising changed so that a decent image of your community can be anticipated from the public regarding this product.

yours truly,

"NEEDLESSLY VULGAR" – Frederick D. Herzog III jump-started his entry into the surfboard wax industry with a brilliant brand name that, while completely meaningless, is alliterative, symmetric, and sounds naughty. An offended citizen of Ventura sent this angry letter to the Goleta Valley Chamber of Commerce in 1984.

John Dahl's management duties at Surf Research included making deliveries. His beat ran from San Diego to Santa Cruz, and one of the stops was a surf shop in Isla Vista owned by Frederick C. Herzog III, aka Zog. Zog was shaping surfboards and running a small shop near the University of California Santa Barbara campus when fate mixed Zog together with a chemist named Nate Skinner. Zog and Nate combined their business and chemistry training and developed a unique, thoroughly modern, high-performance surfboard wax.

"The original formula was tested and refined over a period of a year or so, I believe," Zog said. "You may want to check with Nate about some of these facts, as he may have a better memory than I. This testing occurred in 1969 at my surf shop, which was located on Santa Barbara Airport property. I was shaping, glassing, sanding, and glossing custom boards for local customers and for shipment to some east coast surf shops. I don't remember if it was Nate or I who suggested trying to come up with a surf wax formula, but Nate said he thinks it was my idea. At that time we were doing this for a lark, as I had no intention of ever entering the market with a surf wax. This was when Surf Research was still marketing their products, and I didn't think there was any way that we could be successful against an established company whose owners included some world-famous surfers. Besides all this, surf wax was wholesaling for about 10 cents a bar and I just couldn't imagine selling enough of it to make a living."

Fast forward to the spring of '72. Zog was working at "Project Design," an auto body, custom car prototype shop. This shop was also located on Santa Barbara Airport property. Zog had been working there for two years and wanted to do something else, so he quit. It was at this time that Zoq met with Fred Ryan at the Chart House Restaurant.

"Fred advised me that Surf Research had been sold and was no longer making any wax for the surf market. This was when I actually started to believe that there might be a chance of success for me in the surf wax market. It's ironic, but if Fred had told me that an employee of Surf Research [John Dahl] was going to start his own surf wax company, I might never have gotten started."

In Santa Barbara, Zog found himself in the right place at the right time. There was an opportunity to slip into the market in the transition from Surf Research to Wax Research, and Zog slipped in. The original Sex Wax "factory" was a rickety 20' x 20' garage that Zog's friend, Ken Opiat, sublet for $20 a month. He mixed the ingredients in a 50-gallon drum and used an old boat propeller to stir the wax. Zog used old soup cans for his first molds and then up-scaled to tuna fish cans.

ZOG'S "SHACKTORY" – The original Sex Wax factory in Santa Barbara was this rickety 20' by 20' garage, rented for $20 a month. Zog mixed the ingredients in a 50-gallon drum and used an old boat propeller to stir the wax. He eventually built custom equipment and was in serious surf wax production by mid-1972.

"It took a while to build some custom equipment and get started, but I was making surf wax by June or July of 1972," Zog said. "Our first offerings included peppermint, spearmint, orange, lemon, and banana-scented waxes. Within a short period, these evolved into coconut, strawberry, orange, lemon, and pineapple. This original formula has remained virtually unchanged since its inception in 1972. The only changes that have been made were necessary adjustments to compensate for certain raw materials that were discontinued by some vendors. This formula is not what I would consider cutting-edge, but is still very popular with a significant number of surfers."

Zog had a product and needed a name. He looked to an artist friend, Hank Pitcher, to design a label for the new surf wax. Pitcher came up with a logo bearing the name "Sex Wax," which Pitcher thought sounded phonetically cool: two words of three letters, both ending in "X." Zog liked it too because this was the '70s, and he saw it as a spoof on Madison Avenue's not-so-subtle attempts to use sex to sell a product.

RIGHT TIME – As Surf Research transitioned to Wax Research, Zog moved into the marketplace with Sex Wax. The surf culture loved the outlaw name and bought t-shirts along with wax. His original formula, Zog said, "has remained virtually unchanged since its inception in 1972." Shown here is an early Sex Wax ad from *Surfer* magazine.

FIRST LOOK – Sex Wax bar from 1972 with label featuring the very first graphic design employed in production. Over the years, changes have been made to the graphics and the current label for Sex Wax Original Formula is very similar, but updated.

A BAR IS BORN – Fred Herzog created a monster out of a simple product. Some shop owners were reluctant to carry the surf wax because of the provocative name, but hustle breached the walls. He moved from the Santa Barbara "shacktory" factory to larger production facilities in Goleta in 1975.

Zog loaded up his truck with the first batch of Zog's Sex Wax and drove up and down the coast, trying to convince shops and stores to sell a product with a risky name. Retailers were hesitant at first.

"Initial sales in the first half-year were about $5,000.00, or about 42,000 bars," Zog said. "Second year sales were about 64,000 bars, third year was 120,000, and fourth year was about 240,000 bars. Sales continued to rise very quickly and eventually growth slowed down to a more reasonable rate by the '80s."

By 1975, increased sales allowed Zog to move up to a small warehouse in the heart of Goleta. Wax production took up all the room, so at night Zog would take the unwrapped wax home where the bars were wrapped and boxed for shipping. Meanwhile, the ever-growing demand for Sex Wax t-shirts continued, and much of the inside of Zog's house was stacked floor-to-ceiling with giant boxes crammed full of t-shirts.

The effort was intense but it was working. The business thrived.

"In 1983 we were rather crowded at our Goleta location," Zog said. "So I had been looking around for a somewhat larger facility. The commercial real estate market was very depressed at this time and I was able to purchase an industrial condominium at a great price. As it turned out, my mortgage payments at the new factory in Carpinteria were actually less than the rent I had been paying in Goleta."

"When Zog came in the picture we were worried about competition," Dahl said. "But there were no dirty deeds. We were both on the road at the same time and if there was a dirty deed it was like whoever got into Huntington Beach first would hit every shop and fill the shelves with their product and kind of leave the other guy hanging in the breeze. If Zog was coming south from Seal Beach and I was coming north from Dana Point we would both try to get to HB first and hit all the shops and not leave much room for the other guy's product. Once a shop owner bought from one of us the other didn't sell anything. So it was competitive, but not cutthroat. Zog and I personally did the road for years and years. The battle was who got to what territory first and how much wax. We've always gotten along. I'd say we are the competitive colleagues."

Zog agrees.

"It is interesting to note that although we compete fairly vigorously in the market place, John Dahl and I have always maintained a cordial and at times supportive relationship. I have visited John's facilities on a couple of occasions and John has visited our factory on two or three different occasions. John put me in touch with Stan Pleskunas, who helped us put some equipment together for our production needs. Some time ago I was able to reciprocate by giving John some advice about how to handle one of the more exotic and problematic materials that can be used in the manufacturing process. I have also provided John with some raw materials when he needed them. Our formulas and pricing are proprietary but we share the satisfaction of producing a unique accessory for an extreme sport."

VISUAL – Zog continued the tradition of innovative advertising for surfing products. Since the main job of surfboard wax is to provide traction, the stickier it is the better, like flypaper.

YESTERDAY – Some of Zog's early advertising was—set against today's heightened sensitivity regarding things sexual—startling. So we cropped the ad.

TODAY – Frogs? Why not frogs? They have big sticky feet and fit nicely on the distinctive Sex Wax logo.

SPOOFIN' – With dollar signs in his eyes, Zog made fun of himself and his success in this promotional artwork.

HUMPTY BUMPITY – Quick Humps was Zog's answer to Wax Research's Sticky Bumps. Nate Skinner worked up more than 120 samples to produce just the right textured wax.

Lumps and Bumps and Quick Humps

John Dahl introduced Sticky Bumps in 1992. Surfers went for the new texture, and Zog stepped in to meet the challenge with his own rough-surface wax. He saddled it with a groaner of a name, Quick Humps, which stuck to the Sex Wax theme and suited surfers just fine.

"John Dahl's Sticky Bumps was first on the market, and it definitely raised the competitive bar for surf wax traction," Zog said. "Fortunately, as it turned out, Nate Skinner was familiar with a key ingredient in this new formulation and we were able to start a testing program right away. In reviewing my notes on Nate's Quick Humps lab samples, it turns out that he produced more than 120 different sample formulations before I settled on a series of four I thought would give us an edge. Since then we have added two additional softer formulations, which brings the entire range to six different formulas. My first sale of Quick Humps was July 2, 1997."

Chapter Four

Decisions, Wax, and Hindsight - The Mrs. Palmers Story

By Alistair Cook with Ben Marcus

On any given wave in surfing there is a choice of paths—stall, pull in, hit the lip, accelerate or cut back. The choice of direction is always complex. That moment of split decision determines how well the wave is ridden. In hindsight the choice is easy, but at the time, the choice is not so clear. The surfing industry is no different.

This is the story of two people and the choices they made: Glen Delaney and Al Cook, the founders of Mrs. Palmer's wax brand. They developed the product in Australia in late 1988 as a result of their desire to pursue a lifestyle within the sport they both loved.

They were by no means pioneers in this field. In the U.S. market, there was already an established industry dominated by Mr. Zog's Sex Wax and Wax Research, which made every brand name from Bubble Gum to private labels. Both were established in the early 1970s and survive today in the global surf market.

In the Australian market, the industry was dominated by brands that came and went. It first started with petrol stations such as Ampol marketing paraffin waxes to surfers in the '60s. Products in the early '70s included Honey surf wax and Strawberry Fields surf wax, made by the Ladkins in their Kirra-based establishment. There were plenty before them and plenty after them. For a while, the best surfers like Mark Warren and Nat Young put their names to waxes. Brian Creegan from Ocean and Earth marketed many wax brands through his Sussex Inlet-based accessory business. Survival wax with its boxing kangaroo had a very strong run, as did Summer Breeze. Most of the early waxes were dominated by fantastic smells and their ability to outperform those that went before them.

In the late 1980s came the attack of everything that was green. Slimes dominated the southern market and Green Stuff in the north. These brands were superceded by better products; basically there was not enough money in it to keep the owners pursuing the wax business, so they all went and got real jobs.

It was around this time that Glen "Sloppa" Delaney and Alistair Cook decided they didn't want real jobs and decided to pursue their own business within the surfing industry. Delaney's background was ten years as a production manager at Balin, the '70s surfing leg rope pioneer company based in Flinders, Victoria. As part of his role, he purchased and distributed the leading waxes of the time and sold them to surf stores around the country. With long service leave at the ripe old age of twenty-six, Delaney decided it was time to get out of Flinders and move north to warmer waters.

Around the same time, longtime friend Al Cook had recently returned from an eighteen-month world surfing trip and was desperately looking for any excuse not to go back to the corporate world from which he'd escaped a year and a half earlier. They looked at the market, and at the time it seemed every aspect of the surf market was well accounted for, so they decided that wax was the only product left to do.

Two years earlier, Delaney had done some formulation work on waxes, so he revisited his earlier work, pulled mum's sauce pans out of the cupboard, and began experimentation. They knew that in order to succeed they would have to make something equal to or better than anything available on the market. The work began in late November, 1988. Hundreds of formulations, hours and hours of surfing, and nearly six months later, they finally settled upon the launch formulations. In those six months, they successfully destroyed Cook's mum's kitchen, migrated north to the Gold Coast, and decided on a name for their wax.

Besides the need for a great product, they knew the name and marketing was of vital importance. In addition to surfing, Cook had devoted part of his earlier life to studying business marketing at University. It took him three years, but with his degree came the realization that sex sells. To make matters more challenging, they were entering a market dominated by one of the great marketing names and slogans: "Mr. Zog's Sex Wax – The Best Wax For Your Stick." The naming task was going to be difficult. It is still a subject of mass debate as to what Cook was doing when he came up with the name of the wax. According to Cook, he was pumping petrol at the time. He recalled an earlier survey of how women found pumping petrol an enjoyable sexual activity. He was trying to figure out how he could come up with a name based upon sex and grip, when the old Australian lonely man's best friend, Mrs. Palmer and Her Five Daughters, dawned upon him: "Mrs. Palmers, The Best Grip Around."

"Finally, the missing link between having a good product and a successful product had arrived," Cook said. "We were ready for business."

AUSSIE ANNOUNCEMENT – The founders of Mrs. Palmers surf wax paved the way for their new product with an ad in the June 1989 Australian *Tracks* magazine. Shown, the original hand-lettered, pasted-up ad, loaned courtesy of Alistair Cook. Cook and Glen Delaney spent six months testing formulas before marketing.

A two-story rented house in Miami on Queensland's Gold Coast became the starting point for the Mrs. Palmers business. The longtime friends lived together upstairs and the wax-pouring factory was downstairs: a 44-gallon drum, gas heating element, and benches laden with plastic cups being filled with hot wax. The aroma of wax in the air spread nearly as quickly as the wax on the upstairs carpet. Every day Delaney poured the wax and Cook drove the coast in the Ford Econovan, selling boxes of waxes to surf stores, milk bars, and service stations. The early days were full of fun times, with plenty of parties, hard work, and no money.

"Our savior was Balin," Cook said. "Glen's longtime employer supported us through buying the wax in bulk for distribution to the southern market and paying for it quickly to support our drinking habits. They also gave us the rights to sell their established brand of leg ropes and accessories into the Queensland surf stores, which really got us a foot in the door of a closed surfing market."

By August, the wax had spread to stores throughout the country and well into the floors of the rented house. Besides being lifelong friends and business partners, Delaney and Cook have birthdays that fall on the same day, August 20. This year was no different. The usual party left an even greater mess than was normal at the Miami residence. On the morning of August 21, 1989, the Indian landlords of the Miami home turned up for a family holiday in their downstairs storeroom of the house, as landlords tend to do from time to time.

Cook remembers: "To this day we are not sure if it was the beer cans or the wax on the front lawn, driveway, stairs, kitchen, carpet, etc., but the eviction notice was in the mailbox by lunchtime. It was the calling card we had been waiting for, and time for Mrs. Palmers to get its own premises.

"In September 1989 we moved into the original Mrs. Palmers factory in Nuban Street, Currumbin. It was small, dirty and subject to flooding—but there would be no landlords to rid us of our mess. Glen poured wax into 44-gallon drums out of an old coffee tin with a makeshift tap on it."

WORKIN' IT – Cook and Delaney were not shy about exploiting the bloke joke behind the name Mrs. Palmers. The name, Cook said, "really helped establish [the product] on the grapevine in the Australian underground." They designed wax for warm and cold temperatures. *Courtesy Alistair Cook.*

With the 90s came a new era of business.

"After the first summer, we replaced our manual accounts system with a computer. By mid-1990, business was going well and the business was having no problems paying Glen and me our $100 a week drawings. At the time the dole [unemployment benefits] was paying about $140 a week. If we were ever going to make the wax game work, we would need to go global in order to get the volumes. In June 1990, an American traveling companion of Al's named John Thomas came to visit. We decided to form a three-way venture and set up a wax manufacturing plant on the west coast of the U.S.A. to enter that market."

A plan was formulated whereby Delaney would spend the first six months helping John Thomas set up business in America while Cook ran the Australian business. After six months, they would swap.

"At the time, the first six months in the U.S.A. seemed like a great opportunity for Glen," Cook said. "but the realities and difficulties of setting up business in a foreign land were soon upon us."

Cayucos, just north of Morro Bay, was a small sleepy coastal town and home of John Thomas. Affectionately nick-named "the cereal muncher" throughout his time in Australia, Thomas took Delaney and all of his wax-making abilities into his home for the startup of the Mrs. Palmers business in the U.S.A.

"We had a good formulation in Australia but we soon found out it was going to be difficult to replicate."

To source raw materials, Delaney and Thomas found themselves deep in the 'hood of western LA, which Delaney had heard so many stories about, back in Australia.

"We soon found waxes to be very different in the U.S.A.," Cook said. "And the materials we thought were the same as we were using in Australia turned out to be very different. Instead of a milky brown, one of the key materials was almost chocolate, so we were forced to fill our waxes with whitening agents. This started the trend of the chalky white wax that Sticky Bumps later ran with in following years. In Australia, we had made wax by pouring into plastic cups, which formed part of the packaging at point of sale. Due to the lower wholesale price of 30 cents per bar in the U.S.A., this would not be cost-effective, and we were forced to develop wax molds and shrink-wrap packaging."

Besides draining their limited capital setting up the relevant machinery, Delaney found that the difficulties of getting wax out of molds and then putting it through heat tunnels to package soon became part of his every-day nightmare. It was taking a very long time to make the wax and even longer to package it using this new technology.

"This was not a problem," Cook said, "as it sure was proving even **more** difficult to sell.

"The Mrs. Palmers joke really helped establish the name on the grapevine in the Australian surfing underground. We soon realized that the American sense of humor was a little different. Sure, some of them had grown up with the Rosey Palmer joke, but it was no folklore as it was in Australia. Our ads in *Surfing* magazine soon tried to explain our strange Australian sense of humor through spelling out the fact that Mrs. Palmers had five daughters and was in fact a hand."

FIRST U.S.A. ADVERTISEMENT – Glen Delaney was dispatched to California to oversee the Mrs. Palmers startup operations in the U.S.A. The company's first American ad, shown here, appeared in PSAA magazine in 1991. The unusual brand name puzzled shop owners, who required the Aussie explanation. *Courtesy Alistair Cook.*

The wax pioneers of Zog and Wax Research spent their early days selling wax from vans, and it was to be no different for the small Mrs. Palmers company. Thomas and Delaney drove the length of the California coast, dealing with car breakdowns and a real lack of interest from surf shops. The short stocky Australian and the lanky cereal-munching American looked strange enough, but for most store owners there was an even more confused look on their faces when they saw the Mrs. Palmers name. The standard sales pitch would involve a five-minute explanation on the name followed by an understanding nod but by no means any laughing, let alone buying. The storeowners had Zog's Sex Wax, and the Killer Bees wax was having a good sales run at the time, so they certainly didn't need some Australian joke. In addition to these, Wax Research was making custom-labeled wax for everybody and it seemed with every day came another new brand. On one of the many bad-news phone calls to Cook in Australia, Delaney cited the names of thirty-eight competitors now battling for shelf space in a market where bars wholesaled for 30 cents a bar. The sourcing problems, the manufacturing difficulties, the money problems, a foreign land, and the lack of sales were all becoming too much for Delaney. It was only the afternoon session at the Cayucos tavern that kept him coming back for more each day.

"It must be said," Cook recalls, "that despite a real lack of sales interest, we had incredible support from the local surf stores, Wavelength in Morro Bay, Central Coast Surfboards in San Luis Obispo, and Good Clean Fun in Cayucos. It was these stores that kept us making wax each day. But like Zog's and Wax Research, it was the East Coast that really kept us in business in these early days. In response to our ads in *Surfer* magazine, we got an East Coast rep, and his enthusiasm kept us making wax. Some things never change. He was then and still is our Florida rep, a great guy and a good salesman."

Despite their East Coast rep's enthusiasm and the strong support from local stores, they were getting deeper into debt. Cook's six months in the U.S.A. was nearly up. They thought a West Coast distributor would be their savior. According to the stores they dealt with, Larry Block was the best distributor. A quick call to Block Surf resulted in just another knock-back, and it was back to the tavern for another afternoon session.

In actuality, Delaney had achieved a hell of a lot in his first six months, but the lack of sales and increasing debt had him revealing to Cook in his last phone call from the U.S.A. that it would be fine with him if maybe they just gave up.

Cook was forever the optimist. He had spent the last six months running the comparatively easy Australian business. Delaney's final words were, "Maybe you have just got to come and see how bad it is, but if you want to give up, that's fine with me."

Prior to Delaney's return to Australia, he and Thomas had also encountered a massive amount of problems with the shrink-wrap packaging processes they had tried to introduce for the wax.

"Besides turning out wax that looked the same as the other thirty-eight waxes on the market, our primitive equipment was forever melting the wax, the plastic seal, or both. It was also incredibly slow, so we needed an alternative. Recycling was the new phenomenon in the U.S.A., and we decided upon adopting a recycled paper wrapper concept we had recently seen invented on an Australian competitor's wax, called Green Stuff."

It was to be the first turning point on the road to success in the U.S.A., although at the time, they did not know it.

Cook came to oversee the U.S.A. operations. He had visited John Thomas in Cayucos during his earlier world trip. This trip was a little different. On arrival, the characters in the local street that Delaney had described all came to life. During the wax pouring each afternoon, Earle, a 48-year-old reformed alcoholic and surf maniac, would drop by with a couple of Dr. Peppers for a chat. The neighbor on the other side was a fellow named Dave, a university student who spent most of his time studying and complaining about his girlfriend. He had spent many a session with Delaney at the Cayucos tavern. In the hours following Earle's surfing stories, Dave would inevitably drop by with a beer and the offer of a trip to the tavern.

The wax was getting made in what had previously been John Thomas's backyard garage. They had the capacity to make about 1,000 bars a day, which more often than not would take most of the day. They heated the wax in drums and dispensed it into molds using watering cans. A cooling system built by Thomas and Delaney using an air conditioner was soon obsolete, so they relied on an open garage door and the cool Cayucos weather to help release the wax from the molds.

Sales were still small, but thanks to continued strong support from the East Coast stores, Central Coast Surfboards, Wavelength, and a contract to make Bird skateboard wax, they began to have just enough money to start paying back the large debts that had accumulated while setting up the business.

A real inspiration occurred one day after John Thomas returned from Central Coast Surfboards, where he had just met Dave Letinsky the longtime right-hand man at Zog's Sex Wax.

Dave had gone out of his way to offer Thomas a friendly hand if they ever needed it. He also extended an offer to come down and visit their Santa Barbara factory. The thing that sticks most in Cook's mind about Thomas's meeting with Dave was the news that Zog's was doing 8,000 bars a day.

"This is what reinforced us that maybe there was a reasonable wax market that could make our struggling business viable."

Through the Balin contacts, they secured orders for

France and Portugal, which would give them a small push along. But it was the West Coast of the U.S.A. that was proving to be the biggest problem market. Driving the coast was not cost-effective for sales, and they needed to get another product line to distribute or get a distributor. Balin was tied up with Rip Curl at the time but was getting a bit lost in the Rip Curl machine. For a while, Palmers considered pushing for this line to further establish its own distribution channel. In the end it was decided to give Larry Block one more shot. From out of Thomas's cupboard came an old typewriter, and Cook spent the day preparing an elaborate marketing-style package to try and convince Larry Block to take them on. They mailed it with all five fingers of Mrs. Palmers's hand crossed.

John Thomas sure did like surfing. It was September 1991, and as it usually is on the central coast of California in autumn, the wind was offshore and the surf was particularly good. On returning from a lunch time surf up the coast at Cambria, there was a message on the answering machine from Larry Block. He had received the package and was very interested in distributing the line. This was the turning point that would eventually put Mrs. Palmers on the map.

Block made regular wax orders for at least fifty boxes each week. By this time, the small company had blown up the motors in two trucks by overloading them with wax, so all that was left was Cook's black Toyota Celica hatchback. Each week he would pack wax into every available space in the car—in the hatchback, behind the seats, in the passenger foot well, and on the seats and arm rests. Seventy boxes, or just under a ton, was the record.

Off he'd go on the three-hour drive down to Block's factory in Simi Valley, just north of Los Angeles.

Block took them to the San Diego trade show, set them up with the best distributors in Japan and Europe, and, most importantly for the cash-starved business, he paid on the spot. He is the distributor to this day and the real reason why Mrs. Palmers is still available in most surf stores in the U.S.A.

Cook returned to Australia for Christmas 1991. It was decided to leave Thomas in charge of running the business until Delaney could make it back in late January. Thomas was a great guy, a good business partner, and a very hospitable friend. He was an avid surfer, but working was not really his thing. In the last twelve months he had struggled through the set-up, endured two Australians residing at his house, and had wax throughout garage, lawn, car, driveway, and house. He had inherited enough assets that he did not have to work, and six weeks of running the business solo was enough for him to realize this was not the dream job he thought it would be. By late January he called up Delaney and Cook and said he was keen to sell his share of the business for basically what he had put into it.

With the loss of the U.S.A. factory and partner, it was time for some careful decisions to ensure they did not put themselves back into a position similar to that of the prior twelve months. They had a small but growing demand in the U.S.A., but nowhere to make the product. After careful cost analysis it was decided they had enough volume to cost effectively produce and export it from Australia. Delaney was charged with the job of going to get all the equipment and wax molds from the U.S.A. The paper-wrapper was proving popular in the U.S.A. Block purchased every last bar of wax that Delaney made in the U.S.A. and placed his first export order from Australia for 50,000 bars for shipment ASAP. The real problem was that they only had enough molds in Australia to make 1,000 bars per batch, and the order needed to leave within two to three weeks. Cook remained in Australia, selling by day and pouring wax in the early mornings and evenings to try to fill Block's order. The help of longtime friends and employees of Darren Guest and Paul Smith enabled them to meet these first orders to the U.S.A.

Like anything in running your own business, you solve one problem only to be confronted with the next. Each problem seems worse than the last. In this case, the next one occurred with regard to the formulation. The American business was built on American materials and characterized by the cakey white appearance masking the problem of the brown raw materials. Upon moving the business back to Australia, they started shipping Australian formulations, which were clear in appearance.

Sales soon took a turn for the worse. Competitor John Dahl of Wax Research had seen the sales success of Palmers Wax and was quick to respond with his own version, named Sticky Bumps—a cakey white wax designed to capitalize on the look that many surfers were now seeking. With a little work, Palmers was quick to improve its Australian formulations in line with the former American-made products. But Block was sitting on a lot of stock. Sticky Bumps was moving in on the market they had left behind.

Fortunately for Palmers, its Japanese distributor was in need of wax and was happy with the Australian formulations. Block was able to ship all his stock to Japan, and the newer formula was shipped to Block to start playing catch-up on Sticky Bumps. Zog's Sex Wax soon launched their Quick Humps. The three brands still account for the majority of the wax sales in the world wax market today.

Palmers has since invested in two other Gold Coast factories and still operates from its newly-built Burleigh Heads premises. Delaney and Cook no longer pour wax. They employ a staff of about twenty people in their business selling waxes, deck grips, and, most recently, shoes and clothing. Mrs. Palmers is available in more than twenty countries worldwide and has re-launched itself with its latest innovation of wax sold in a cardboard box that doubles as a storage container.

Cook summarizes the entrepreneurial journey:

"This story is just the Palmers's side of the difficulties encountered in the set-up of a wax business. With both

John Dahl and Zog making wax for nearly twice as long as Palmers, I am sure they both have their own tales to tell. As wax makers, we share a common bond and often will talk at various trade shows around the surfing industry.

"In recent years, people have often asked about the large volume of waxes we each must be manufacturing on a daily basis. At Palmers, we still consider wax as one of our core lines but it is by no means where we make the money. The current surfing industry is a long way from our beginnings back in 1989 and even further away from the '70s when John Dahl and Zog's began. To survive in this world of publicly listed surfing companies where every Quiksilver, Billabong, and Rip Curl want to put their names on every single surfing product to maximize their shareholders' profits, small companies do not have the option to be highly specialized and must get bigger. The surfing industry has become an ocean, and its eat-or-be-eaten mentality means that companies such as Palmers have been forced to pursue growth strategies to stay in business. In recent years at Palmers, we have sold a lot of our comedy line of clothing and also expanded into the footwear business through the launch of the Kustom brand. [Recently] we imported more than twenty containers of footwear and clothing, and it helped us come to the understanding of who chose the right path in the surf industry.

"If you take your standard forty-foot container, it holds about six thousand pairs of footwear or twenty thousand pairs of boardshorts. In wholesale terms to the surf store, a forty foot container of these products is worth approximately $500,000. To reach this same dollar in wax at wholesale, you would need to sell a million bars of wax. Wax is a specialized field and it cannot be made in China, so the margins are also nowhere near as good as the apparel industry. The history of the wax business shows hundreds of people have tried to put their name to wax, but the bottom line is you need to be doing in excess of a million bars a year just to make it viable. It takes years of experimentation and development of specialized equipment to make it. It is also a labor-intensive and messy business, which has kept it a bit of a specialized business you only get involved in for love of the sport and the lifestyle.

"I think the key thing that probably Zog, John Dahl, and ourselves have in common is we all deal with wax in and out of our factories by the ton and occasionally by the container load. It arrives in bulk by forklift and leaves in a smaller bar, but also by forklift on pallets and sometimes containers. We often find ourselves noting it comes in by the ton, goes out by the ton, but it never comes back.

"We could all answer the question of how many bars we produce a day but no one can answer the question, 'Where does it all go?' Rubbish bins, old boards, carparks, deep into carseats, back into the sand, or back into the ocean or earth where these materials initially came from. No one really knows. At Palmers, the key we have learned after fifteen years of hard work is that someone must be getting a hell of a lot of good surfs and great waves in many different parts of the globe.

"Unfortunately for the big three wax makers, while we are helping to make these surfs more enjoyable for most surfers, we probably aren't getting our fair share of the waves. We have been too busy dealing with making it and running our businesses."

Chapter Five

Other Notable Wax Pioneers

Randy Dowler and Waxx-On/Waxx-Off

In 1982, a UC San Diego chemistry student named Randy Dowler entered the surf wax market with a product called Waxx On. There were other interesting efforts at what modern marketing experts now call "branding": Potted Puke and Sick Pig wax from England, Wizz wax "Don't even think of eating this," Toe Jam wax, Slime wax, Ganja surf wax and Skip Frye's Man wax.

WAXX-ON AD – Randy Dowler, a chemist and the inventor of Waxx-On, is the son of a rocket scientist and grandson of a chemist. "There are no surf wax formulas that I cannot imitate, duplicate or copycat," he said. He sold surf wax and accessories to shops along the California coast for more than two decades. Shown, ad from *Surfing*, June 1984.

Dowler Tells His Story to Jefferson Wagner

My great-grandfather, Mr. Dubbs, was the father of my grandmother Ruth Dowler. I was granted my middle name, Dubbs, in honor of Mr. Dubbs, who was educated as a chemist and earned a living as an organic chemist.

My father, Warren Leroy Dowler, was also a chemist. He is a retired rocket scientist with a degree in chemical engineering. He worked for Cal Tech and the Jet Propulsion Laboratory in Pasadena. He designed and built rocket motors—that put food on the table and clothes on our backs. We moved around a lot as his work demanded, but my first years were all spent near the ocean. I grew up in and around La Jolla. I learned to body surf at La Jolla Cove when I was four, and considered myself to be an expert after catching my first wave all the way to shore. Or so I thought.

I got my own chemistry set when I was five and immediately proceeded to burn down the backyard.

My great-aunt owned a house and two vacant lots right in front of the Windansea parking lot, before trading it in for the quiet solitude of a cattle ranch in El Cajon. When I was a kid, my family used to gather 'round the Thanksgiving table of relatives living in San Diego to eat turkey, mashed potatoes, and cranberry sauce while watching the surf break at Windansea.

In 1959, my family moved to San Jose and I looked forward to our trips to Santa Cruz. I liked to go body surfing in Santa Cruz at the mouth of the San Lorenzo River, located just south of the boardwalk along Main Beach.

Nobody owned a wetsuit in 1959—or maybe only a few people. To stay warm, you rubbed Vaseline all over your body, which worked as an insulator against the frigid 55-degree water.

In those days, you bought Vaseline at gas stations, along with other things. One of the products the gas station sold was bars of paraffin wax, because paraffin is one of the first byproducts of the oil refining process. Gas stations sold the wax for practical uses like canning and candles, but surfers used it for traction. I didn't have a surfboard then, but I bought one of those bars for 50 cents. I offered to trade it to a surfer who was on the beach sitting on his homemade surfboard in exchange for a ride on his board. He smiled and laughed. He said, "Kid, if you can carry it to the water, you can ride it." I instantly ran back and grabbed my best friend Gary, who helped me drag the 10-foot-plus log to the water's edge. Together we pushed it out until we could barely stand up, somehow got the thing pointed back to the beach, then belly-paddled it in the soup towards shore. The wave picked us up and we belly rode the thing tandem until it pearled. We came up with sand and saltwater flowing out of every orifice of our bodies. We did that two more times, and pearled two more times. Finally, we caught a wave tandem and belly rode it all the way until the skeg dragged in the sand. We never managed to stand up. I vividly remember that the tips of our fingers had turned purple. We were so cold. But I instantly knew the very first time I pearled on the surfboard, what I wanted to do with the rest of my life. Surf.

Back in the '60s and '70s it was a popular pastime for surfers to build their own surfboards. Since I had hands-on practical experience fiberglassing the family cabin cruiser with my dad, during the summer of 1959 my buddies and I shaped and fiberglassed our own surfboards in high school using the tools in my dad's workshop.

During high school, my father built his own black and white TV and then waited for the first broadcast station to come on the airwaves. He later built his own color TV and also waited for the broadcast station to broadcast in color. We were the first people on our block to own a microwave oven.

My father was and remains an avid amateur astronomer. Working together, we designed and built telescopes for personal use on top of Mt. Wilson. Incidentally, my father was on the team of scientists who were at the Mt. Wilson Observatory that recently re-silvered the mirror on the 200-inch reflective telescope. I grew up with magazines like *Sky and Telescope, Science, Scientific American* and numerous *Chemical Journals* scattered across the family coffee table in front of what my father referred to as "the Boob Toob." I guess he didn't like what the networks finally started broadcasting.

To make a telescope mirror, one casts a wax blank that is impregnated with Carborundum [evenly sized sand particles], which are used to grind and shape the glass blank before adding silver to its surface to create a highly reflective mirror.

On a lark, I created my own surf wax formula from the wax materials that were left over from grinding an eight-inch telescope mirror. With a genetic family background predisposed to an interest in the field of chemistry, I whipped up a quick formula off the top of my head and passed it out to all my surf buddies to get their opinion. It worked.

In the mid-70s I graduated from the University of California at San Diego with a degree in physical chemistry and a minor in fine art. Chemistry is the study of how to make your own nuclear weapons, laser beams, and particle accelerators for fun and profit. Chemists just love to create and destroy things, all in the same day. It's a trip. For two summers and part time during the college school year, I worked for the Scripps Institute of Oceanography while surfing La Jolla Shores or Black's Beach before work, during my lunch hour, and after work.

In the youth of my college days, Black's Beach, La Jolla Shores, Windansea, and Bird Rock were my regular surfing stomping grounds. During college, I lived on Electric Street in La Jolla for two years, a couple of blocks up the street from Windansea, before moving to Del Mar.

Just after graduating from college, I moved next door to one of my surf buddies in Del Mar, Hal Handley, who lived next door to Julie Herzog, Rick Herzog's sister. One of Hal's best friends married Julie Herzog and eventually became a very successful science fiction writer. I purchased one of the very first available Sex Wax Ts from Julie Herzog, but I was never an avid fan of his surf wax formula. I stuck to my own knitting.

During the '70s I had been making batches of surf wax, off and on, just for fun, while building surfboards professionally for several north San Diego County board builders: Clinton, Channin, Town & Country, Donald Takayama, Nectar, Lightning Bolt, Agua, Encinitas Surfboards, etc. Also during that time, I was hanging out at a beach break south of the small Mexican fishing village known as Puerto Escondido, the "Mexican Pipeline." There, a well-known traveling Hawaiian surfer walked up to me on the beach and asked me why my surf wax was not melting off my surfboard, as his was. I handed him a bar of my surf wax formula while wolfing down the last of my banana-macadamia nut pancakes. He tried it. The surf wax, that is. Later that night, I got a knock on the door of my hotel room at the Rincon del Pacifico, which charged $1.50 per night, split between two surfers. The Hawaiian surfer came asking for more. His buddies were standing in back of him for moral support. I passed it out free of charge. For the rest of the trip, he bought all the banana-macadamia nut pancakes that I could eat for breakfast. Life is good.

When I returned home there was a postcard from South Africa, requesting that I ship two cases of surf wax to his Hawaiian address. I whipped up the formula, shipped the wax, and forgot about it. Two months later I got a check in the mail and a request for more surf wax. Soon I was getting mail from all over the globe, requesting that I ship people surf wax, with money inside the envelope to pay for it. Wow! Lucky me.

I sold a surf wax tool under the brand name Waxx Off—the double XXs in honor of my great-grandfather Dubbs. The surf shop retailer's standard reaction was, "Why would somebody want to purchase a wax tool? Why not just use sand or put it out in the sun?" That wax tool later evolved into a popular product generically known as a wax comb. I was the first to market a wax comb. I also whipped up a batch of surf wax and passed it out for free to surfers in surf shops along the sales route, as a promotion for Waxx Off. I asked Jack Briggs, a surfboard salesman for Russell Surfboards in Newport Beach, what I should call the stuff. Since I was selling Waxx Off to remove and shape surf wax, he suggested "Waxx On." What else? Waxx On gets it on, Waxx Off gets it off. Done deal.

I approached Rick Herzog of Sex Wax, Inc., in Goleta with marketing the product, Waxx Off. He turned me down.

When I had obtained purchase orders for one hundred cases of Waxx On, I borrowed the money to set up a surf wax factory, then delivered the goods C.O.D. It was all downhill from there. I never intended to become a surf wax manufacturer. It all just happened. During my first year of commercial operations, I was shipping surf wax to Japan, Australia, the East Coast, and South Africa. All the advertising was strictly word of mouth. They came to me—all I did was make the stuff. There was no advertising budget. I didn't have enough money to advertise.

During the early 1980s, a Hollywood scriptwriter who was an avid Malibu surfer incorporated the product names of Waxx Off and Waxx On into a movie script, *The Karate Kid*. My national advertising campaign was piggy backed into a Hollywood blockbuster. I never paid a single penny for a very successful national advertising campaign: Waxx On/Waxx Off.

Also back in the early '80s, Zuma Jay had a dinky surf shop located a quarter-mile down and across Pacific Coast Highway from the Malibu Pier, and he sold lots of surf wax during hurricane season. Zuma Jay, Encinitas Surfboards, Newport Surf 'n Sport, and Kennedy Surfboards were among the first surf shops to receive commercial shipments. Waxx On surf wax turned out to be a very popular item, but its brand acceptance by today's marketing standards would be considered slow. In fact, Waxx On basecoat was the very first basecoat invented. I coined the term. It was a hard sell to retailers. Of course, it didn't hurt that a surfer by the name of Gerry Lopez took a shine to the wax formula.

By 1983, sales volume for my surf accessories business was getting a little more than I could handle all by myself. I needed the assistance of a competent secretary to handle the telephone, take surf wax orders, answer mail, ship U.P.S., and spend all my money. Since I didn't have a lot of money to hire an experienced secretary, I married a gal by the name of Karen Lee Stonehouse and immediately put her to work in the surf wax factory. On our honeymoon, we drove up the coast of California collecting past due accounts from Johnny-Come-Lately surf shop owners. She still to this day performs a lot of secretarial chores in my day-to-day business operations, and she still spends all my money. Lucky me!

During the early- to- mid-1980s, a kid by the name of Tom Curren worked behind the counter of a tiny surf shop called Channel Islands. He told me that he liked Waxx On, but asked if I could make the formula a bit more sticky. I said yes, and returned with a sample formula that eventually became known as Kangaroo Goo. It later turned out to be Tom Curren's favorite surf wax. At the time, Tom was the best-known professional surfer of the day, the Tiger Woods of the surfing community. With Tom's input helping to design a stickier surf wax product for his own personal use, he promoted Kangaroo Goo wherever he went—for free. The product eventually climbed in popularity to become one of the most popular surf waxes in California during the 1980s. However, about ten years later, the creation and popularity of Sticky Bumps during the early 1990s eventually knocked it off the shelf.

During the late 1970s, a few years before introducing Waxx On to the commercial marketplace, I toyed with a formula that is similar to Sticky Bumps. It had nearly identical components but not the same ratio of identical materials. Unlike Sticky Bumps and Quick Humps, which sinks like a rock in salt water, [my product] Air Wax floats like a cork. It received so much negative response from the majority of surfers who tried it that I never made the decision to market the product until the late '90s, under the brand name Air Wax. It rips. In terms of stickiness, it's not as sticky as Superglue™ but it will definitely rip out all your pubic hair the first time you attempt to stand up. Surfers have a love/hate relationship with the Air Wax formula—you either love the stickiness or you hate it. Retailers constantly whined and sniveled that it was too sticky. Personally, in cold water, nobody does it better than Air Wax, in fact, not even half as good.

In response to retailer complaints about the product being too sticky, I toned it down with the introduction of both the Dr. Sticky Fingers and The Good Stuff surf wax formulas.

All brand name products have a natural life cycles of birth, growth, maturity, and decline. Waxx On Basecoat is now the oldest surf wax formula still in existence on California surf shop shelves, having outlived all the other surf wax formulas that were popular in their day, including the Original Sex Wax Formula.

Koala Bar, Bad Wax, Mondo Surf Wax, and a list of other wax formulas were soon to follow. By 1985, the Waxx On, Kangaroo Goo, and Koala Bar brands were household names in the California surfing community. I had inadvertently captured 70% of the surf wax market share in California, before moving the surf wax factory north to Cambria, a small coastal village on the tail end of Big Sur. The Wax Research brand, so popular in the 1970s, had declined. Sales of the formula for the original Sex Wax in California had fallen to less than 30%. Some surf shops in California even stopped carrying the brand.

Bubble Gum surf wax was my chief competitor during the 1980s. While Bubble Gum surf wax went for the hoopla—sponsoring lots of surfers to promote their brand—the majority of them were secretly riding Kangaroo Goo or Koala Bar on the sly. Professional surfers who had Bubble Gum stickers plastered all over their surfboards would walk up to me asking for a bar of one of my brands of surf wax, telling me it was their favorite. I asked them, "Aren't you sponsored by Bubble Gum?" They'd say, "Sure, but we like your surf wax better. It beads well and stays sticky for a long time." Oh, well. As Mark Twain once quipped, "There's many a slip, twixt the cup and the lip."

Since I am a chemist by education, there are no surf wax formulas that I cannot imitate, duplicate, or copycat. And since I built my own surf wax factory from scratch, there are no surf wax shapes that I cannot imitate, duplicate, or copycat. The Sticky Bumps and Sex Wax Quick Humps formulas may be highly guarded secrets in the eyes of John Dahl, the president of Wax Research, Inc., or Frederick Charles Herzog III, the president of Sex Wax Inc., but not for yours truly. I even know the formula for Mrs. Palmer's. It took me all of five minutes to duplicate the formula in the wax lab. I also know how to produce formulas for skate and snow wax. It's not rocket science.

My original purpose in manufacturing wax was to develop a long-term business relationship with surf shop owners to get my foot in the door so that I could sell other more profitable items like clothing. I suppose I did that, although others sold the products I helped to develop.

For most manufacturing products, the total average annual advertising budget is 2% of gross annual sales volume. At 4% you are definitely over-budget and are burning money faster than it is coming in. The total annual sales volume for Sticky Bumps is $350,000, according to court documents placed in the public record by WRI. At 2% of $350,000, that leaves an annual gross advertising operating budget of just $7,000 per year. If you call up *Surfer* and *Surfing* magazines and ask them to mail you an advertising rate card, you will soon discover that $7,000 per year doesn't buy very much advertising. When it comes to advertising in the surf magazines, all that glitters is not gold.

The vast majority of surf accessories companies have gone over the falls financially by over-advertising, by giving in to the temptation to burn money faster than it is coming in. The name of the game in business never changes—buy low, sell high!

Surfing consumers make their purchasing decisions based primarily upon formula. Surf shop retailers base their purchasing decisions primarily upon low price. Advertising is hideously expensive and is only highly effective when done in large doses of monotonous repetition. Large advertising budgets are mutually exclusive with selling at low prices. If you drop into the surf wax market with a large advertising budget, then you have committed yourself to riding a close-out all the way down the line with zero opportunity to kick out. Previous to starting my commercial production of surf wax with the Waxx On brand, I worked for a lot of surfboard manufacturers who bet the farm on large advertising budgets and then bit the bullet.

G&S Surfboards was at one time the largest surfboard manufacturer in the world. After reaching the Mt. Everest of surfboard production, they bet the farm on a combination of large advertising budgets and low prices. The results were quite predictable. They went over the financial falls, big time. Bubble Gum surf wax, Killer Beez, On A Mission, Wet Dreams surf accessories—the list is nearly endless.

The market forces in surf accessories, including currently popular surf wax formulas, are extremely fickle. What goes up often comes down. You can be riding high in April and shot down in May. Most surfers and retailers measure success by total sales volume or momentary popularity. That is an extremely shallow yardstick by which to measure a person's contribution to the world of surfing.

Brand name popularity comes and goes as surely as the sun rises in the east and sets in the west. When it comes to popularity in brands of surf wax, you should think in dog years. Sticky Bumps is twelve years old. Is that young or old in dog years? And you already know what they say about teaching old dogs new tricks.

Never a horse that couldn't be rode,
Never a rider that couldn't be throw'd.

After 23-plus years of loading up the van with surf wax and other surf accessories and going door to door along the California coast selling to surf shop owners, I have lived a happy life.

People ask me if, at age 52, I still surf. The answer is, "Why not?" After a surf trip to Kauai in 1993, complications from long-term effects of surfer's ear caused my ears to seal shut and I could not hear for three weeks. To solve the problem, I drove out to the desert. In the low humidity, my ears finally cleared and once again produced audible sound. From that experience, I radically toned down my surf time, and turned to fly fishing for fun and to give my ears a needed rest from the ill effects of over-surfing. A few months later, I ran into a guy at Morro Bay who told me that when he moved back east and stopped surfing, his surfer's ears healed up.

I decided to cut way back on surfing rather than endure surgery, which produces long-term hearing loss from the drilling noise. And surfer's ear comes back a few years later anyway. If you undergo the surgery enough times, you will eventually go deaf. In 1994, I had an ear doctor check my ears, and he informed me that it was the worst case of surfer's ear he had ever seen. He was amazed that I could still hear.

To prevent myself from being overly tempted to surf every time the surf got good, I intentionally pulled the plug on Waxx On and wisely decided to voluntarily downsize my surf accessories business, including the surf wax production at the factory. I simultaneously decided to expand my economic horizons beyond the surfing industry. In 1994, I seriously got back into fly fishing in the Big Sur River. I was hopelessly smitten with steelhead fever. The cure is worse than the disease.

That same year I moved to Oregon. My surfer's ear eventually healed up and the bone growth in both of my ears has dramatically receded over time. I still surf, but not to my heart's content. I limit my surf time and always wear a hood. I was never able to adjust to earplugs, as it ruins my balance. So I surf less and avoid getting my ears wet in cold water, while exposing them to the ill effects of cold air. A lot of time during the winter, I wear a hood when fly fishing to protect my ears. It works.

I no longer consider myself to be a surfer per se, rather, I have always thought of myself in terms of being an avid waterman. Surfing, body surfing, water and snow skiing, motor boating, sailing, diving, kayaking, rowing a drift boat, fishing, panning for gold—all are water-related sports.

I currently write a weekly fishing column for both the *Daily Triplicate* in Crescent City, California, and the *Curry Coastal Pilot* here in Brookings, Oregon.

Can I tell you a secret? Wherever you happen to find a run of steelhead on the west coast of the Pacific, there is usually an excellent surf break located not far away.

My greatest accomplishment in life? That's easy. Becoming a grandfather.

"If I have seen further, it is by standing on the shoulders of giants." —Sir Isaac Newton

Happiness in life is not about getting what you want. Happiness in life is all about wanting what you actually get once you have it.

Fifty percent of the fun in surfing is while you are paddling back out, watching some guy get sucked over the falls, chopped up in the meat grinder, and dragged across the inside reef while his lungs are screaming for oxygen. The same goes for surf wax manufacturing. I rode Waxx On, Kangaroo Goo, and Koala Bar all the way to the beach until the skeg was dragging in the sand. Then I walked up the beach and sat down. Simple enough.

Con Colburn, Innovator

Constantine "Con" Colburn, who opened the Surf House surf shop in Santa Monica in 1958, created a line of surfboards and accessories known for quality and cutting-edge ideas. As he grew from a ding-repair shop to the manufacturer of the innovative blunt-nosed "Ugly," the "Super Ugly," and "C.C. Rider" surfboards, he joined the surf wax competition. His brand became hugely popular.

Colburn promoted his Control (or Con Trol) Competition Wax as "the ultimate in board control in all types of surf conditions." In the 1970s and '80s, Colburn was one of the first to sell massive quantities of surf wax from his factory in Venice. He sold his surfboard wax through Standard Brands Paints stores and was a pioneer in the production of private labels. The warm yellow bar with the private label Superwax is the rarest of all significant surfboard waxes and can be found only in a few collections.

His Control Wax remover—another radical idea—was the companion product for Control wax. Colburn also marketed a surfboard traction spray.

Colburn retired to Bishop, California, far from the waves, and died at the age of 57 in 1992. His company name has been brought back to life by new owners who shape boards with the same famous monikers. But the new Con is not in the wax manufacturing business.

GOLDEN ERA – Con Colburn placed his indelible stamp on the early development of surfing accessories and products. In this wax and wax remover advertisement from *Surfing*, June 1969, he declared that the paraffin a surfer's grandmother used to put up preserves was "ancient history."

CONTROL PRODUCTS – Colburn's wax, the ad says, lasts a long time, won't rub off, is tacky in warm and cold water, won't melt in the sun, and "won't attract the slippery ocean residues," which doesn't say much for the health of most ocean waters. Colburn's direct, no-frills approach appealed to serious surfers.

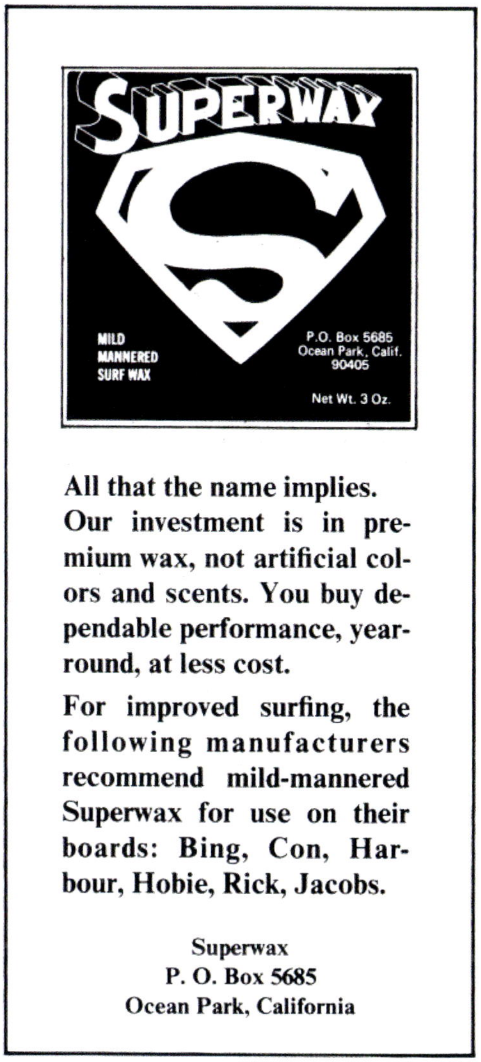

MILD-MANNERED SURF WAX – No silly colors and scents for these dudes. Con Colburn pioneered the private label approach and was the first to mass market surf wax. His company was located in Santa Monica. Con's waxes are rare finds for the collector.

Bubble Gum Comes and Goes and Comes Back - The Galland Brothers

After an association with John Dahl in the early 1980s, brothers Britt and Grant Galland of Carlsbad, California, began producing Bubble Gum surfboard wax out of a garage. The product was naturally pink, and they added a bubble gum scent. Jim Watson and Eric Baker were part of the company. In 1986, they licensed the name to "Life's A Beach," aka Bad Boy Club.

In 1989, the Gallands licensed the Bubble Gum name to Michael Tomson for Gotcha. The name was there until 1996 when it went dormant and wax sales ended. Grant Galland was in surf equipment sales management until 2003, when he brought the new Bubble Gum Wax back into production.

According to an article in a business publication dated October 1991, the Gallands had neglected to trademark the Bubble Gum name outside the United States, and found that they had to buy it back from a businessman who trademarked the name in Europe.

"Those were expensive lessons," Britt Galland is quoted as saying in the article.

The Gallands entered the surf accessory business. Today, his new and complete focus, Grant Galland said, "is on the Bubble Gum Wax only. I am establishing a growing base of customers for my private label and secret formula, which is carefully guarded and distributed nationally. Bubble Gum is the #1 best seller out of Waxlab domestically."

RESURGENCE – Grant Galland has revived the Bubble Gum label he and his brother Britt started in the early 1980s. They licensed the name to several other companies, watched it go dormant, then brought it back to life in 2003.

Chapter Six

Traction Alternatives

In issue five of *Surfer*, World Surfing Products of Hialeah, Florida, was offering Competition Surf Wax in six colors at three bars for $1.00, along with a Factory Hot Wax Kit, which was not for bikini lines or an El Camino. World Surfing Products also offered Slipcheck by Morey-Pope, which brings up an overlooked facet of the whole traction game. In the first few years of the desperate movement away from Parowax and mom's other products, it was spray-on traction in a can that got more of a marketing push than wax. First came Slipcheck, developed by Morey-Pope, and then in true copycat fashion there appeared Control and Grip spray and SprayMate, a product from Surf Research.

WORLD SURFING PRODUCTS
TEN EIGHTY-FOUR EAST FORTY-SEVENTH STREET, HIALEAH, FLORIDA ▫ OFFICE 681-5481

Competition Surf Wax
6 Colors 3 for 1.00

Factory Hot Wax Kit
For a Better Job 2.95

Factory Ding Repair Kit
Professional Type 4.95

Factory Ding Repair Kit
With Clark Foam 8.50

Factory Skeg Replacement Kit
Glass and Wood — All Shapes 14.95

Tahitian Print T-Shirts
Screened All Around 2.95

GEORGIA, ALABAMA, FLORIDA AND S. CAROLINA
DISTRIBUTORS FOR:

Slipcheck by Morey Pope
Rincon & Aloah Car Racks

WRITE DEPT. S, 1084 E. 47th ST.,
HIALEAH, FLORIDA.
ENCLOSE CASH, CHECK OR MONEY ORDER,
OR SEE YOUR BETTER SURFSHOP.
DEALER INQUIRIES INVITED.
SEND 35¢ FOR POSTAGE AND HANDLING

EAST COAST MARKETING – World Surfing Products of Florida distributed Morey-Pope's radical new traction idea, Slipcheck. After Tom Morey's board almost sank under the weight of saturated pool table felt, he developed Slipcheck, a variation of Ventura County's yellow traffic barricade paint.

The Traction Revolution - Morey-Pope's Slipcheck

By Tom Morey with Ben Marcus (continued)

One day, circa 1957, it occurred to me that maybe there might be something better than wax, some way to apply permanent traction to the deck. I started thinking about what might do the trick, and decided on fabric. I shopped for an hour or so at a fabric store in Venice and came out with enough green pool table felt to do the deck of my board.

Later that night, I removed all the wax, cleaned it thoroughly with gasoline, then established masked borders around the deck a couple of inches or so above the edge of the rail. I applied contact cement to the back side of the fabric and to the deck and glued this big piece of fabric on there. It laid down really flat and after I'd trimmed off the excess with a razor blade it looked like a really cool, curved pool table.

Stoked, I ditched out of work at Douglas (aircraft) the following morning and drove to Malibu. The surf was chest high. I ran down the little cliff there in the pit area and launched into the water on this thing and MAN oh MAN! Traction you couldn't believe!

Now paddling out, I seemed to notice the board maybe weighing a bit more...but hey, no matter! I get out about half way to the flagpole lineup and along comes a set...first three waves, too crowded...then wave number four has only got two guys already on it. You know how it goes at Malibu—a fella spends the whole day there waiting for a break, so shoulder-hopping is standard practice, even back then. Thus, I swing around, take two strokes, up, fade, turn: and eowweee, I'm like standing on living room carpet. But, #$%^, the two guys BOTH fly past like I'm standing still! Great traction, sure, but HUGE drag, and by now maybe five pounds of extra weight. And after a few more waves on this stuff, I've learned all I want about this stuff and my enthusiasm is over. Sure the traction is super, but now I've got the worst case of tit tears and belly rash you ever saw!

"Excuse me, kid...ahhhh, could you please help me carry this board up to my car? This thing's reallllly heavy and I'm not feeling too good."

The Nose Riding Contest: Traction to the Forefront

Fast forward to 1965. I am no longer a rider on someone else's surf team—having moved through Goodart, Con, Sweet, Velzy-Jacobs and Weber—but have my own growing brand (Morey) and shop (Australian Surf Shop) in Ventura where I've moved to live closer to Rincon. I've been in this business for around a year and am struggling. Not many potential customers up there. So, in order to draw attention to my brand, I stage a surfing contest: The Tom Morey Invitational, Saturday, July 4th. This is the first actual professional contest, there being real cash prizes: $1,500 total. This contest is unique in that it's all about being on the nose. The time is determined by judges viewing through binoculars and collected by their assistants using stopwatches.

But how are they going to see what is the rear limit of the nose—which I've arbitrarily designated as the front 25%—of each board?

Idea!

"Spray that front quarter with a brightly colored paint."

And this we do.

This works out great and the contest the following day is a huge success. But what works best in the water is something two San Diego rider/board makers show up with. Mike Hynson and Skip Frye have pre-measured their boards and applied 3M adhesive-backed black traction pads to their noses. You know it today, of course, as what almost every skateboard in the world is covered with. Back then it was just used on steps and certain patches of airplane wings where someone might step.

Hynson and Frye's whole deal is terrific. They surf great, coming in among the top three. The shapes of their boards are terrific and trend-setting—much shorter, saucy and with a square nose…precursors of the shortboard era. The black traction pads show up clearly. Plus, these traction pads give them extremely firm footing. All quite enviable.

Slipcheck is Born!

During the following week, with all the pros having left town…excitement has died out and now it's life in Ventura as usual. So to overcome the boredom, Bill "Blinky" Hubina and his pal, Bill Delaney, the sander, are secretly pushing two new boards of their own through the shop. Around noon on Friday, Bob Thompson, the glosser, puts on a couple of clear finish coats. Blinky and Delaney go sit on the railroad tracks facing the back of the building…sitting with their back to the sun, waiting for the resin to dry.

Now as fate would have it, at 856 East Front Street, right next to door to our shop at 858, is Art's Barricades, a paint shop that performs most of the Ventura City Street Department's paint work. Day in and day out, Art and his guys are back of his place leaning street barricades up against the back wall and spraying them bright yellow. As the result the entire wall is covered with bright yellow over-spray.

As Blinky and Delaney are dallying back there staring at this wall and yakking, a great idea dawns.

"Look," pointing to the wall. "Let's get Art to paint the noses of our boards so they look like THAT. Bright YELLOW! Bitchin'!"

So they mask off the noses of the now dried boards, haul 'em over there and, "Sure, no problem," says Art.

Zip, zap, sizzzzzzzzzzz, and in a matter of a couple of minutes, both sport beautifully colored noses, just like the pros had. Double bitchin'!

Tech Luck

A brief technical note: This particular paint is extremely tough. In fact, it's the same stuff used to paint yellow lines on streets. However, if it is sprayed from a little too far away, instead of turning out smooth, it comes out with a light granular texture. However, nobody at the moment is paying any attention to this, one way or the other. It's only about getting a FREE paint job that colorful, so: "Great, and thanks!" And the two Bills rush off to Stanley's.

The sun's about to sink when they reach the now-extinct surf spot called Stanley's Diner—named, as are most surf spots, after the most prominent landmark…in this case an actual diner on the old Highway 101 named "Stanley's." As everyone usually does before waxing new boards, they first toss them into the soup to wet and cool the decks. As they retrieve the boards, a check of the "feel" of the new paint jobs is in order, and "Damn!!" they feel great!! Instead of being a smooth yellow, the surfaces are flat and give a distinct traction when wet!!!

They start waxing from the tail forward and since they're in a hurry anyhow, when they get to the noses they stop. Art's worker, who apparently did the job from a little too far back, has taken the traction revolution one step further. An historic choice is made at this point. They decide to not wax over these paint jobs, rather, to just leave 'em and get out there!

Well, this turns out to be a really great surf session, especially with the added traction up front; both guys are giving nose riding its ALL!

Delaney comes up with a name for the stuff right on the spot: Slipcheck.

Marketing Slipcheck

I'm working late at the shop that evening when Bill and Bill come back from Stanley's, really excited! They explain and add that their plan is to borrow some money, buy a generator set and paint spray kit so they can go to Stanley's on weekends and start spraying noses. I suggest that instead they let me have a crack at marketing this stuff…that the name Slipcheck is great and that I'm having an idea of my own.

That idea also turns out to be a good one: get on the phone Monday and shop for a paint-spray canner who could maybe package this thick concoction in aerosol form.

"Let me market it for you guys," I tell them. "I'll write

up the ads, Karl [Pope, my new business partner] and I will sell in quantities to all our surf shops [about fifty], we'll advertise in *Surfer* magazine and pay you 50% royalty on every can sold."

This was a fair deal all around and took them about a half minute to think over.

"Agreed!"

A handshake follows. No paper, no ink, no paranoia. We didn't even know a lawyer.

As it turned out, we were able to find someone who could can it. I copied all the directions and warnings from other spray cans and incorporated them into our label: a cheap black and white job which had a pretty cool homie design on it that Jolly, my first wife, designed.

Then, without permission from OSHA or the EPA or anyone like that—because they didn't exist yet—we just went ahead and started using and then selling the stuff: putting it on a few of our own boards at first, then selling to Tom Hale Surfboards just up the street, Owl Surfboards in Summerland, and so forth.

Meanwhile, all of our emerging Morey-Pope boards—Peck Penetrator, Cooper Blue Machine—now came with Slipcheck included as part of the original equipment. We also applied special Slipcheck jobs to a score of other boards, dolling up used boards for pennies. At the same time, we are selling other surf shops on each of the various ideas, photographing cool, checkered patterns and running ads in *Surfer* magazine, whose advertising prices in those days made sense.

HISTORY LESSON – Endorsing the value of invention, Tom Morey and Karl Pope developed a line of new products for the young surfing industry, including a transportable three-piece board, the snub-design board, and the offspring of traffic signs, Slipcheck sprayable wax replacement.

ON THE NOSE – An intriguing, persuasive photo touts the benefits of Slipcheck and leaves the reader wondering—"Well, *did* he?" Photographer Bill Delaney is one of the "Bills" whose epiphany at Art's Barricades in Ventura led to the development of Slipcheck.

White and Black

Now check this out, because Art's Barricades was still right next door, with the back of the building totally covered with bright yellow over-spray blotches, I was certain that if we put Slipcheck out in yellow, immediately someone else would figure out what this "special" paint was and rip us off. Because there are far more white lines on roads than yellow, I reasoned that whoever was the paint supplier—and Art instantly told me: Baurer Paint in LA—that they would also have white. I asked and certainly they had white. So that's all we sold at first, dozens of cases of white.

Even though white is immediately dirty-looking and the glare from a white nose is tough on your eyes, still, white Slipcheck sold because it really worked.. Come to think of it, just like white zinc oxide, which is ugly as » ^!#!º% ^ & sells, for protection of your human nose.

But dozens of cases were not going to get us anywhere.

One day, I found myself in some nowhere town and noticed that the city had blacked out a few white lines on the street. They'd used black spray paint. I got out of the car, went over and checked the newly painted blackness and sure enough, there was the same bitchin' Slipcheck texture. I dashed to a phone and called Baurer.

"Sure, we sell this stuff in black."

Huge secret smile, low low tech R&D breakthrough!

Sales quadrupled; hundreds of cases moved. Meanwhile, we're increasing the advertising. And regularly, our dear bookkeeper Eileen Velvick figures all this stuff out and gives Bill and Bill their royalty checks.

Color TV Versus Black and White

Although sales are still nothing of much importance, they are enough so I can now afford a color TV. And wow, what a difference that color makes.

So, one day I am down at Baurer's picking up some raw material for the aerosol packager and notice that they have huge barrels of glass beads, many different sizes. These are the beads that are mixed with paints to cause headlights to reflect off various signs and surfaces at night. Beads that can be mixed with paint!

I get a few sample pounds of likely sizes, take them along on a pow wow with the packager and sure enough he can easily make us a paint with one particular size of these beads in ANY color we want! I think about the difference my color TV is making and am going nuts inside with anticipation.

Water testing proves the beads WORK! Well, fairly good, anyhow. Not quite as good as the regular street paint, but because the colors are so good nobody really cares. Both kinds of paint are a lot better than wax: 100% dependable, hot or cold water.

More ads. Sales soar.

Now we are shipping hundreds of CASES per month!

BIG PRODUCT – Sales quadrupled after Morey's "low, low tech R&D breakthrough" inspired by street line paint. This late-60s advertisement for Slipcheck shone with hip copy—"Big texture means big traction in deep soup." By then it came in seven colors.

Granma's Touch

One day, one of the regular groms, Petey Robertson, brings in his grandmother's lace-patterned tablecloth and asks if we'll spray his board through it. We give it a try and it's pretty cool. But nothing smashing. However, based on all the great Mexican black velvet paintings that tourists coming out of Tijuana buy at the border, we try color sprayed through the lace on top of a black Slipcheck field. Results stupendous!

Everyone in the shop goes nuts and now we're spray painting the noses of everything anybody can get hold of. This, too, we advertised and by now, thousands and thousands and thousands of cases are selling to hundreds of surf shops up and down both coasts.

I have never had so much success with anything in my life. Not then, or even ten years later with the boogie board.

And regular as clockwork, Hubina and Delaney got their royalties. OK, maybe sometimes they had to nag a little, but you know, we did pretty good with it.

Breaking Waves

By 1969, drugs had swung through Ventura just like through most of the rest of America. Everything is disrupted. Marijuana, then LSD. One glasser is stoned and although he puts the right amount of catalyst in the resin he forgets to stir it. Parts of the board are burning up. The rest is a soggy mess.

Guys are standing around waving fingers in the air and watching the after-images instead of doing jobs. And our judgment is weakening.

Simultaneously, the shortboard revolution is taking place. Australian Nat Young beats the world's unofficial greatest nose rider, David Nuuhiwa, in the 1966 championships in San Diego, and with this defeat, boards start getting radically shorter. Goodbye nose! Sure the shortboards still had pointed front ends, but now the fronts were becoming something more like a beak than a nose—narrow and pointed—and few were the riders of what was twenty years later to be named a "longboard."

Slipcheck sales die in a matter of weeks. Hundreds of surf shops are stuck with thousands of cases. I quit Morey-Pope and move to Hawaii. Like the stock market crash of '29, this is now The Great Longboard Crash of '69.

The Phoenix Rising: El Gripo

Fast-forward about thirty years to the late '90s and we find a retro longboard movement underway. Whereas the longboard sales of the '60s numbered short of 100,000 per year, shortboards are now selling in the high hundreds of thousands per year... maybe even a million a year (and only Grubby knows). Still, like baby turtles, a lucky portion of shortboarders grow up. And when they do, being heavier and having real jobs that steal their muscles, the ones who don't quit surfing go to larger, warmer platforms on which to sun, paddle easier, and catch waves sooner. Thus, longboards also are being purchased by the hundreds of thousands. Many of the old brands are now re-emerging.

So, starting around 1998, I too find myself back in the surfboard business, making the long ones out of my garage and hankering for real Slipcheck. Especially wanting black on the nose so there is less glare from the sun, same as football players who use black smudges under their eyes.

During those lapsed years, the original suppliers moved on while different traction approaches emerged; some just came and went. I tried crinkle paint: nope, too slippery. I hear about Ultra Deck, an excellent light rubbery paint, but no, they've been shut down by the EPA.

But "Seek and ye shall find," and sure enough, there is a packaging company that's taken up the challenge and so we now market clear and a wide range of colors under our new name, "El Gripo."

NOT-A-WAX – Among other entrants into the alternative traction sweepstakes was the Grip brand, distributed by World Surfing Products. It claimed to "provide the surfer with 42% more gripping power."

So how are sales? Well, not really all that great, with today's advertising costs so high and users tending to watch TV and get tattooed rather than fixing up their boards themselves in the garage. Yet not all that bad either, since there is a real need for a good spray traction and El Gripo (now as an epoxy) really is a lot better than Slipcheck ever was. And many other practical uses have come to the forefront: diving boards, places on boats and around the house or pool, patio, and kitchen—areas that have gotten slippery are easily made safe again. It has many useful places on sailboards, sail and powerboats, jet skis, kite boards, and skateboards.

Time will tell just how much juice is left in this wave.

Astrodeck - the Herbie Fletcher Traction Action

Surf wax marketing from the 1970s into the '80s became almost non-existent. Through *Surfer* issue 18, number 1, Wax Research was advertising surf wax in six different formulations: Hot Climate was 72 degrees and above, Summer was 64-72 degrees, Cool Water was 56-66 degrees, Winter was 60 degrees plus, Soft was 50-70 degrees and Sticky was 55-75 degrees. They were also offering Fruity Flavors: Tangy Orange, Lime Green, Red Cherry. But those ads faded almost to zero into the '80s. Wax Research and Zog's apparently decided that form of marketing was not necessary. There were ads for other waxes here and there—for Heritage Surfboard Wax out of New Jersey, Bolt Wax from Hawaii, Control Products wax, Wombat Wax out of Glendora, and Honey Bee Wax from Wave Crest Hawaii.

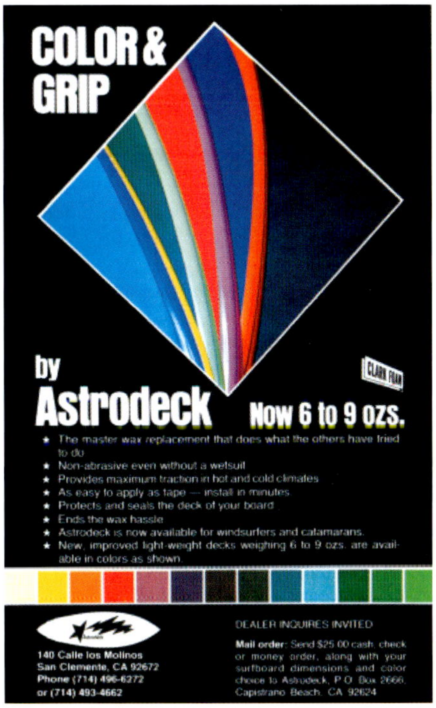

COLOR AND GRIP – Advertisements for Astrodeck touted the product as "the master wax replacement." It was, said the ad, easy to apply and also protected and sealed board decks. At $25.00 a pop, the product was pricier than a 25-cent cake of wax.

FRESH IDEA – In 1976, Jim Van Vleck walked into Herbie Fletcher's Dana Point surf shop with a new traction idea, a liquid polyurethane elastemer. Recognizing its value, Fletcher marketed it as Astrodeck, or Astro Deck. (The spelling was fluid.)

As the '70s faded into the '80s, the ads for alternative traction methods dominated the pages of *Surfer*.

Herbie Fletcher was to the '80s what Mike Doyle was to the '60s. Herbie was the huckster—he was everywhere, selling surfboards and clothes and most importantly, selling traction products. Astrodeck started an advertising blitz that made a lot of ad guys happy through the '80s. Astrodeck was traction pads. Ultradeck was spray-on traction. Their campaigns were some of the most consistent, dominant ads from the '80s into the '90s, featuring just about every hot surfer on the planet. It appeared as if they had forsaken wax for Astrodeck or its Australian competitor, Gorilla Grip. Surfers were decorating the tails and noses of their boards with the traction material developed by Herbie.

Zog commented on both the threat posed by Astrodeck and other traction ads and the changes taking place at his company: "As regards the traction pads and other products which promised to replace surf wax in the '80s, these products and their ad campaigns created a certain amount of concern, but our sales were still growing throughout that period. I took a 'wait and see' attitude regarding the need to respond or counter those advertisements."

The truth is, wax was selling itself and both Zog and Dahl were busy organizing and reorganizing their businesses. Zog relocated his factory operation to larger quarters in Carpinteria.

"When Surf Research moved to Denver," Dahl said, "Waxmate went from being the number one seller to disappeared in about a year. In the meantime, Sex Wax along with Wax Research's brand and custom label shop brands took over the market. Bubble Gum came along in '83 and was a strong brand through the early 90s."

Herbie's Story, with Ben Marcus

Herbie Fletcher is one of the pillars of the surf-industrial complex, right up there with Greg Noll and Mike Doyle and Hobie Alter—all men who combined their love of surfing with their desire to produce, create and enable.

"Let's start with the basics. The entry. Surfing was just fantastic when I was nine years old and I just wanted to go surf because it was a fantasy. I was on vacation in San Clemente down at T Street and grabbed a surfboard when somebody left it and went and paddled. Then I went home and had a paper route...

"Home was Pasadena at the time. I had a paper route and I bought a #27 balsa wood surfboard. So I started on balsa wood and in those days we used candles or paraffin. We'd heat it up and do hot-waxes. And sometimes we'd use a brush so it made bumps and we'd get bumps. So that was cool and then I moved to the beach, to Huntington. And your wax job said a lot. In those days you didn't have wetsuits and it was cold, you know?"

The year was 1957. Fletcher was in San Clemente on vacation with his parents.

"I never thought I was going to be a surfer but I liked the beach. I liked the ocean. I liked the water and liked to *play*. I really liked to play, and it takes a kid who really likes to play to go surfing. So I wanted to go surfing. This was way before the magazine came out but we had heard about some of these guys, these surfers.

"And so I was hanging at Huntington. By then the magazine came out and I just wanted to meet some of these surfers. I didn't know anything about surfing. But hanging around surfers and hanging around good surfers, you get into it and you watch them and you want to do what they're doing and so you study their styles. My mentors were Phil Edwards, Miki Dora, Dewey Weber, and a few others we'll get into. I imitated those guys but I developed my own style. It's just like anything. You watch people and study them to get going and then you apply yourself to it and get going."

Fletcher's family moved to Huntington Beach, AKA Surf City.

"Yeah, I got lucky. They loved me. So we moved to Huntington and I was surfing and being stoked and then Hobie and all those guys...I wanted to be more like those guys because they were shaping and making surfboards. I wanted to hang around the shop. I wanted to learn everything about surfboard making and surfing. And that was my life. I loved it. That was my passion.

"So Hobie gave me a job in about '65 doing a clothing line with him. I was the youngest guy and I made enough money to go to Hawaii and live for a while. And enough money to get back. While I was over there I worked and did odd jobs and did whatever I could to stay there, living in empty cars, vacant cars or houses, and the bushes—wherever I could stay. And that was it. I quit school and lived there. That was what I loved. I wanted surf and so I was working with Phil Edwards and these guys down at the Hobie Shop in the summertime, learning everything about shaping and glassing and buffing out and all that.

"Then Slipcheck came around and that was really bitchin'. That was in the mid-60s, '66, right around there.

"Before surf wax…there was paraffin and stuff to make candles with. Anyway, I got into Slipcheck. And other people had been trying to do stuff because we'd been out at Sunset surfing, and when you're out in those big waves and just using paraffin… the water is warm in Hawaii, so it had some good grip, but it still would flatten out and rub off and get slick, especially if it was raining with the muddy water. So there was always a concern for developing better traction for your surfboard. But the Slipcheck just didn't have it because it was too abrasive on you.

"So moving along into the '70s—I don't know when they came out with all these different waxes. I'd just get whatever I could and even when they were making Wax Research, it wasn't really special. It was better than paraffin but it wasn't real special yet.

"So this guy came by my surf shop in Dana Point and had this traction system and it was Astrodeck and he poured it on surfboards."

The man was Jim Van Vleck. It was 1976. Van Vleck did not stay on the scene.

"He just sort of faded away. He wasn't interested in it. But I was really interested because I knew it worked. Gerry Lopez knew it worked. Mark Richards knew it worked. Cheyne Horan knew it worked. Shaun Tomson. All the greatest surfers in the world knew that it worked."

It was not a spray-on system like Slipcheck.

"No. This was a material for the deck of your board. Not wax. Not a spray-on. I had developed a tape to go on the back so we had full decks, and then I did little cut-ups. I was doing advertising in magazines with full decks and nobody caught on. Not the surfers. Not the surf shops. I had it in surf shops for a year and couldn't sell it, so then I went, 'Okay, I am going to do this last ad. If it works, great. If it doesn't work, I'm over it.'

"But you've got to realize this whole thing was out of love. It wasn't out of the money. I'm in love with surfing.

"So I put these little pink and black squares in *Surfer* magazine. A half-page ad or something like that and the punkers went off on it. They loved it. Punk was in and Newport went off and it spread. It made enough money to pay the rent and get going. It's never been a big money deal and I've always done it out of a need. I get to work with all the greatest surfers in the world and play with them. And I'm a shaper. I shape surfboards and I design them. I began on balsa boards and went through the '60s and competed at the Duke and I rode big waves and the Pipeline, so I was always working on surfboards and shaping them and laying up fin panels and cutting them out and doing things with surfboards.

The advertising blitz did not start until the 80s.

"We did some advertising in the '70s but it didn't go anywhere. I was ready to close it down but I didn't want to know about it. But then we did that small ad, and that small ad helped to pay the rent. In my shaping room, I would shape in the morning. I had these molds and I

SPRAY-ON ALTERNATIVES – Spotted on a shelf at Herbie Fletcher's workshop were two old cans of the biggest spray-on traction products, Morey-Pope's Slipcheck and Fletcher's Ultradeck.

would pour Astrodeck in the afternoon. The fumes would be so radical we would have to get out of there quick.

Fletcher put all the raw material together.

"We were making our own. We don't use it now, so [I can say what it was]. Yes, it was an elastemer. A polyurethane elastemer. It was originally made for roofing or tracks—like what the Olympic guys run on. I was working with sailboarders in the '80s and making a bump in the center of the board like the Arch Bar is now.

"They needed a solid grip on their booms, so I made Boom Grip. Then I put this little center thing in the center of their boards for their back foot so they could have leverage and really push. I made Pro Strips for Shaun Tomson and he was putting a Pro Strip on top of a Pro Strip and it would make a little arch in the center of the board. Now we have this arch and we're developing kicktails to have a place to put your heel with your arch and your toes. We started cutting little holes through it to give your toes a place to go down into these holes and really get a good grip. So they're just *tearing* the lip off the waves now."

Wax was still being used along with Fletcher's additions.

"They're still using wax in the middle. A lot of guys were using front foot pads and so I developed the Arch Bar, which is a common thing. We had the kicktail with the holes, which were called Monkey Grip. All this had developed because I am a surfer, but I am also a skater and I was taking my kid to skateboard parks. I knew what was next in the evolution to surfing was flying in the air. All these guys were trying to get air but they didn't know how to land it. I was taking my son to the skate park so he could come in and land these airs in the transition with these front-foot pads and the tail pad. I was making hook for training so they could have hooks on their back foot so they could hook in, and I was doing all these crazy things because I really wanted to see people fly in the air and do flips and 360s and barrel rolls and this crazy stuff that hadn't been done yet. I was always one to push the limits because I had learned from such great guys and they could talk to me. Anyway, I developed it for Jet Skis and sailboats and baseball bats and hammers and martial arts. I was developing it for them but it didn't sell."

Astrodeck did sell well to sailboarders and snowboarders.

In the '80s, *Surfer* ads show that a sizeable portion of the great surfers of the world used Astrodeck or its primary competitor, Gorilla Grip.

"The team was so large and so big and so good," Fletcher said, "it was everybody who could really surf well in the whole industry. I had all of those riders on my team because there was one common thing: they could ride Astrodeck and it didn't offend any of the clothing companies or the surfboard companies because that is who was sponsoring people in those days. Everybody could ride under the Astrodeck label, so we used to have these great team photos."

Group shots of Fletcher and the surfers he sponsored, boards at the ready, are like a Who's Who of surfing, displaying a special tribe, a golden commune. It was a colorful, dedicated fraternity. But the camaraderie did not translate into monster monetary gain.

"It [Astrodeck] never made a lot of money. You're dealing with the tail of a surfboard, so that is why most people thought it was crazy to do this when all the clothing companies went huge. And they go, 'Well, why are you doing this?' And I go, 'Well, this is my little thing and I have a real good time playing with all the surfers.' People couldn't understand that: the love of surfing. You look at all these people who do all these other things in the industry. Do you see them out surfing? Do you see them playing on surfboards? Do you see them designing surfboards? The other companies are just companies. They are out for the dough and I am in it for the love. I'm just having a good time still playing with my surfboard and I'm still surfing Pipeline. I'm twice as old as the oldest guy out there, pretty much."

Fletcher's attempt to adapt Astrodeck to Jet Skis never panned out.

"I developed it heavy for Jet Skis and WaveRunners and all the personal watercraft, but the people in that think it's too expensive and makes all their stuff look bad."

Astrodeck's heyday, Fletcher said, was "probably around 1991 to 1995."

Fletcher employed thirty people during those years.

"Probably thirty, but we were making everything here. And now it's mainly sales. But I still design it all. I just made a new pad for Gerry Lopez."

The very first Astrodeck ad in *Surfer* magazine featured Gerry Lopez.

"It's been all this time. He's been a rider since. He's probably the first rider you know beside myself. Mark Richards was in there. Rabbit. Bertlemann. I just made him a new pad and it's awesome. It's camouflage with a lightning bolt going through the "A" and he's just all stoked."

MAN AT WORK – Herbie Fletcher still designs the Astrodeck pads. Here he fabricates a custom piece. "This is my little thing and I have a real good time playing…" he said. In the 1980s, the Astrodeck team was sizeable because the product did not infringe on other major sponsors.

Chapter Seven
True Wax Tales

The Subject is Sexism

In Australia and in warehouses scattered all over Southern California, there were various attempts to horn in on the surf wax market dominated by Zog's and Wax Research...and the accent is on the word "horn." Mrs. Palmer's and Her Five Daughters wax made its first appearance in the April 1991 issue of *Surfer*, but there were other attempts.

Wet Dreams surf wax did not last too long.

Big Peckers surf wax offered free key rings.

And who can forget the Poontang's ad campaign, which had a long run of half-pagers featuring a lot of moist tongues and up-thrust cleavage that bordered on soft-core porn?

Even the innocuous Bubble Gum wax stooped to using the "Blow Me" slogan. That slogan and the Bubble Gum brand was an open invitation for kids to put in their mouths what they should be putting on their surfboards. Although every bar of Bubble Gum wax came with a "Do Not Eat or Chew" warning on the label, there were incidents of junior high kids around the country getting sick and going home early from surf wax eating sessions.

Through the '90s, Zog's ramped up their product line and their marketing, offering sun screen and wax remover and taking out full-page ads in *Surfer* that showed a little creativity going back to the Mike Doyle days. One ad explained that Zog's might be a little too sticky, and showed a bar of wax with flies stuck to it.

In the late '90s, Kelly Slater came out with K Wax to go along with K Grip and K Every Else, but that wax, like the other waxes, eventually disappeared. Wet Dreams dried up, the sales of Big Peckers drooped, and Poontang's finally slithered away, leaving only Mrs. Palmer's to become the third of the Big Three that would dominate the surf market into the twenty-first century.

Slapdown by Viagra®

The combination of Zog's successful Sex Wax label and the target market of testosterone-rich young surfers still prompts wax makers to go to great lengths to find tempting, impulse-item names. One Huntington Beach entrepreneur filed a trademark application to register the name VYAGRA. The VYAGRA brand wax was already being sold at the time, and the intent to take advantage of the famed sex-enhancement pharmaceutical product was all too obvious. A stern "Cease and Desist" letter from the real Viagra's® law firm quoted from the wax packaging in pointing out the trademark infringement— "Get Up!" "Stay Up!" and "Without VYAGRA, Consider Yourself Soft."

Stop using the VYAGRA name immediately, the letter said, and then we'll talk about our client's claims against you.

The surfboard wax makers backed down.

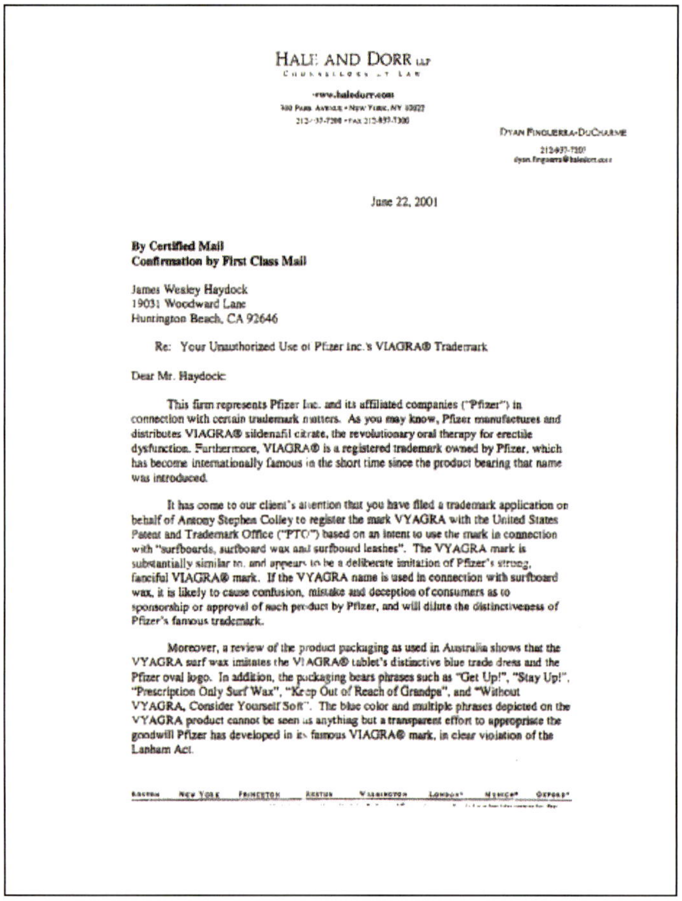

HALE AND DORR LLP
COUNSELLORS AT LAW
www.haledorr.com

DYAN FINGUERRA-DUCHARME

June 22, 2001

By Certified Mail
Confirmation by First Class Mail

James Wesley Haydock
19031 Woodward Lane
Huntington Beach, CA 92646

Re: Your Unauthorized Use of Pfizer Inc.'s VIAGRA® Trademark

Dear Mr. Haydock:

This firm represents Pfizer Inc. and its affiliated companies ("Pfizer") in connection with certain trademark matters. As you may know, Pfizer manufactures and distributes VIAGRA® sildenafil citrate, the revolutionary oral therapy for erectile dysfunction. Furthermore, VIAGRA® is a registered trademark owned by Pfizer, which has become internationally famous in the short time since the product bearing that name was introduced.

It has come to our client's attention that you have filed a trademark application on behalf of Antony Stephen Colley to register the mark VYAGRA with the United States Patent and Trademark Office ("PTO") based on an intent to use the mark in connection with "surfboards, surfboard wax and surfboard leashes". The VYAGRA mark is substantially similar to, and appears to be a deliberate imitation of Pfizer's strong, fanciful VIAGRA® mark. If the VYAGRA name is used in connection with surfboard wax, it is likely to cause confusion, mistake and deception of consumers as to sponsorship or approval of such product by Pfizer, and will dilute the distinctiveness of Pfizer's famous trademark.

Moreover, a review of the product packaging as used in Australia shows that the VYAGRA surf wax imitates the VIAGRA® tablet's distinctive blue trade dress and the Pfizer oval logo. In addition, the packaging bears phrases such as "Get Up!", "Stay Up!", "Prescription Only Surf Wax", "Keep Out of Reach of Grandpa", and "Without VYAGRA, Consider Yourself Soft". The blue color and multiple phrases depicted on the VYAGRA product cannot be seen as anything but a transparent effort to appropriate the goodwill Pfizer has developed in its famous VIAGRA® mark, in clear violation of the Lanham Act.

BOSTON NEW YORK PRINCETON RESTON WASHINGTON LONDON* MUNICH* OXFORD*

MAXWAX Pat on the Back

Jefferson Wagner
Zuma Jay Surfboards
22775 Pacific Coast Hwy.
Malibu, Ca. 90265

Dear Jefferson,

On a recent surf trip, I went into your shop in Malibu to buy some lo cut booties (cut foot). I was totally impressed with your surf wax collection and would like to add to it. I own Surfwax Candle Company. We make a surf wax called MAXWAX. I have enclosed one of each of our three temperatures to add to your collection. Our product is sold primarily in Florida on both coasts, but can be found from Texas to France.

I also understand you will be coming out with a book. I would really like to buy a copy when it comes out, so save my info.

Thanks,

Steve Willis
Surfwax candle company
MAXWAX

SURF WAX FAN – Like Rick's Café in Casablanca, everybody comes to Zuma Jay's in Malibu, where visitors can buy or rent fine surfboards, wetsuits, and kayaks and choose from racks of California beachwear. Jay has one of the top wax collections in the world.

Mr. Haydock
June 22, 2001
Page 2

Pfizer has asked us to contact you in an effort to resolve this matter promptly and amicably. Please provide us (by facsimile to the above number) with your written assurance that you will immediately

1. expressly abandon the VYAGRA trademark application with the PTO;
2. cease the development, manufacture and sale or distribution of any and all products which use the VYAGRA name;
3. cancel any further plans it may have to develop goods or advertisements incorporating the designation VYAGRA or other colorable imitations of the VIAGRA® trademark, including the advertisements posted on the <www.surfing-life.co.uk> website; and
4. desist from any other infringement, dilution, or unauthorized use of the VIAGRA® trademark or color blue imitations thereof, including use of the VYAGRA mark.

Pfizer has spent considerable sums of money and effort developing its valuable trademark rights in its VIAGRA® mark and is committed to diligently prosecuting and enforcing those rights. Once we have your written assurance of compliance with the above demands, we will be in a position to discuss an appropriate settlement of Pfizer's claims against you.

Nothing herein is intended by us, nor should it be construed by you as a waiver or relinquishment of any rights or remedies which our client may have in this matter, and all such rights and remedies are hereby specifically reserved.

Very truly yours,

Dyan Finguerra-DuCharme

cc: Tiffany Trunko, Esq.

CEASE AND DESIST – Pfizer, Inc., maker of the sex-enhancement drug Viagra®, took exception to a none-too-subtle campaign by a surfboard wax company to hijack the product image, issuing this strongly worded demand to stop immediately. The VYAGRA wax company backed down, retiring the name and the innuendo-heavy slogans such as "Get Up!" and "Keep Out Of Reach of Grandpa."

Sharkbait Wax - Fin? What Fin?

Notes from co-creator Dan Argabright

Sharkbaitsurfwax is the brainchild of a couple of friends from Southern California, Bruce Condon from Culver City and Dan Argabright of San Clemente.

"The idea for Sharkbaitsurfwax," Argabright said, "emerged while I was out surfing. I would imagine giant sharks swimming beneath me. It dawned on me that when you paddle out to go surfing, whether you liked it or not, you were sharkbait.

"The notion of sharkbait and surfing stayed in the back of my mind. I wondered how could I connect the two in a way that symbolized the courage of the average surfer who says, 'Yeah, I know there're sharks out there, but I'm going out anyway.'

"Then one day Bruce said, 'How about Sharkbaitsurfwax?' Only later did we realize the perfection of the concept. It didn't take long to visualize the perfect logo for Sharkbaitsurfwax: the trademark silhouette of a surfer holding his board, out of which is a huge shark-bite—a surfer whose body language says, 'Get me another board. I'm going back out!'

"With the help of our surfer friends, we immediately began developing and testing a high-quality surfboard wax, one that would live up to the name Sharkbaitsurfwax.

The rest is going to be history."

UNDAUNTED – The Sharkbaitsurfwax celebrates surfers' fortitude in the face of the real danger of lurking sharks. A good day surfing, it conveys, is worth taking the chance. Dan Argabright of San Clemente and Bruce Condon of Culver City own the company.

Friction Testing: Surf Wax as a Science Project

Surfboard wax is all about friction. He who discovers the perfect coefficient of friction between board and boot wins.

Friction is a resistive force that prevents two objects from sliding freely against each other. The coefficient of friction (**fr**) is a number that is the ratio of the resistive force of friction (**Fr**) divided by the normal or perpendicular force (**N**) pushing the objects together. It is represented by the equation: **fr = Fr/N.** [School for Champions website]

Friction is the surface resistance to relative motion, as of a body sliding or rolling [Webster].

Friction is resistance offered to the movement of one body past another body with which it is in contact. The amount of friction depends on the nature of the contact surfaces and on the magnitude of the force pressing the two bodies together, but not on the surface area of the contact surface. [Concise Columbia Encyclopedia]

The different types of friction are static, kinetic, deformation, molecular and rolling. Each has its own coefficient of friction.

The symbol in physics for the coefficient of friction is the Greek letter mu, μ.

Scientists have determined the coefficient of friction for various surface combinations—steel on steel (dry), steel on steel (greasy), Teflon® on steel, brake lining on cast iron, rubber tires on dry pavement, metal on ice, and rubber tip of crutch on rough wood. Lubricated and smooth materials have a very low coefficient of friction.

The search for the wax with the most effective coefficient of friction is a serious quest for consumers as well as manufacturers. To discover the best of three waxes, one high school student applied the scientific method in his California State Science Fair project.

CALIFORNIA STATE SCIENCE FAIR
2003 PROJECT SUMMARY

Name(s)	Project Number
Kurt A. Miller	**S0214**

Project Title

Surf Wax Friction

Abstract

Objectives/Goals

A problem that most surfers have is what surf wax they should use and which wax performs the best. In acknowledging this problem, I decided to test multiple waxes to determine which one performs the best under different conditions. I decided that I should test Sex Wax, Terra Wax, and Sticky Bumps as these surf waxes are the most popular in San Diego. The wax that performs the best should enable the surfer to perform sharp maneuvers on the board's side without allowing any slippage or loss of grip. Understanding this, I decided to test the height at which the foot could remain suspended without slipping down the face of the waxed board.

Methods/Materials

I defined the best performing wax by the condition that yielded the greatest height at which the "foot" was able to stay suspended without slipping. During each trial, there were some variables that might have effected the performance. The temperature of the air was 62 degrees Fahrenheit and the humidity was around 80%. The weather conditions may have effected the performance of each wax. After testing the waxes multiple times, each wax showed smoothing of the surface of the wax from repeated use. The only reason that the wax performance may have been different under wet and dry conditions was the wax gradually began to get smoother.

Results

Based on the product test, if I were to choose a surf wax, it would be Terra Wax. I would choose Terra Wax for multiple reasons. Terra Wax has treeless paper for its wrapper, which biodegrades quickly, is non-toxic and doesn't hurt the environment. Terra Wax performed best in my test as well.

Conclusions/Discussion

After collecting and analyzing the data, I conclude that wax B-Terra Wax, out performed wax A-Sex Wax and wax C-Sticky Bumps. Terra Wax out performed the two other waxes by achieving the highest average height of the board before slippage occurred. In the actual surfing environment, this result would correlate to better adherence of the surfer's foot to the board in conditions of steeper angles.

□After testing, I have determined that my hypothesis was incorrect. I found that Terra Wax achieved the highest average height before the weight began to slip compared to Sex Wax and Sticky Bumps. Terra Wax also is the most versatile wax because it performs well in all types of water conditions.

Summary Statement

Project is about comparing surf wax friction

Help Received

Advice from Vic Miller

Ap2/03

WAX TESTING – High school student Kurt Miller studied three surfboard waxes for his California State Science Fair project. The winning product was Terra Wax, the eco-friendly packaging of which—in addition to performance—was a significant factor.

More Surfing Camaraderie

Tom Morey on Finding the Good Spots

"One of the beauties of the sport back in those days was the dynamics of friendships. There being no surf cams, surf reports and few phones—certainly no phones where there was surf—you'd depend on past knowledge, a feel for weather conditions, a tip you got at school, a hunch. Since everyone had their own hunches, after a while friends who did and didn't surf at the same relative level sorted themselves out. Guys who were good surfers tended to travel pretty much alone, showing up at the better spots at the better times. Those who weren't so good frequented crummier places."

Surf Wax Is Not Food

(So don't eat it)

The delicious, food-flavored scents and appealing texture—similar to the wax red lips and buck teeth youngsters munch on at Halloween—make surfboard wax a temptation to eat. But the ingredients, especially early on, are hardly conducive to human consumption. A letter to *Surfer*, April 1990, asked the "Surf Docs" if it is safe to eat surf wax. The answer is an emphatic "no!"

Dear Surf Docs

Every time I purchase a bar of wax, I notice a warning the surf manufacturers put on it: "WARNING: Do not eat or chew." Why not? It tastes great and it's less filling. I've been chewing ever since I started surfing, and nothing happens to me.

Dr. Jekyll

Dear Dr. Jekyll

Do you like chewing on dirty chemicals? Because that's what you put in your mouth every time you succumb to the surf-wax munchies. We assume you're not swallowing the wax, which would really be bad. But you should also know you absorb chemicals very efficiently, even just chewing, both through your gums and when you swallow saliva.

The basic ingredient of surf wax is paraffin wax, a synthetic substance, but the type used in surf wax is usually food grade (meaning it can be used in or on foods). For instance, the same type of wax is used in a candy-bar additive, and to put a shine on supermarket apples. Beeswax, or a beeswax-like substance, is also added to give the wax pliability. Some of the dyes are the same as those used to color soft drinks.

Thus endeth the (possibly) food-grade part of surf wax. Now we get to the Toxic Avenger goodies. All surf wax is softened by petroleum-derived products such as Vaseline so we don't lose our nipples when we paddle out. Bottom line: you're sucking on a thicker, slightly cleaner grade of motor oil there, Dr. Jekyll.

Also, surf-wax scents don't come from vanilla beans, bananas, or mint leaves, they come from factories. They're totally synthetic. In addition, some waxes have plastic in them for temperature control. Chalk up another petroleum-derived ingredient.

Finally, synthetic rubber or petro-chemical resin is added to increase the stickiness factor. John Dahl of Wax Research reckons that of all the ingredients, these two might be the most unhealthy to ingest. The fact is, though, nobody really knows about the long-term effects of chewing or eating surf wax because it's only been around for about 35 years. That sounds like a long time, until you realize it took the medical establishment nearly 50 years to prove cigarettes were coffin-nails.

In addition to not wanting you to eat these chemicals, wax-makers like Zog of Sex Wax fame put those warnings on their products because the machines used to make them aren't necessarily clean. They don't have to be. Surf wax isn't a food.

So just because nothing's happened to you yet doesn't mean it won't if you keep chewing wax. Anyway, why eat more chemicals than we already do? You might find yourself turning into Mr. Hyde…and he died an ugly death.

Chapter Eight

Surf Wax, Today's Market - 40,000 Bars a Day and Counting

INDUSTRIALIZATION – The production method for little bars of surfboard wax has come a long way since the days of the garage entrepreneurs. Photographer Jeff Divine shot pictures of the production process at a large, modern surfboard wax factory in Southern California. At 44,000 square feet, the facility covers more than an acre.

The process begins with the raw material, the "essential compound."

Shelves are lined with scents to be added to the wax.

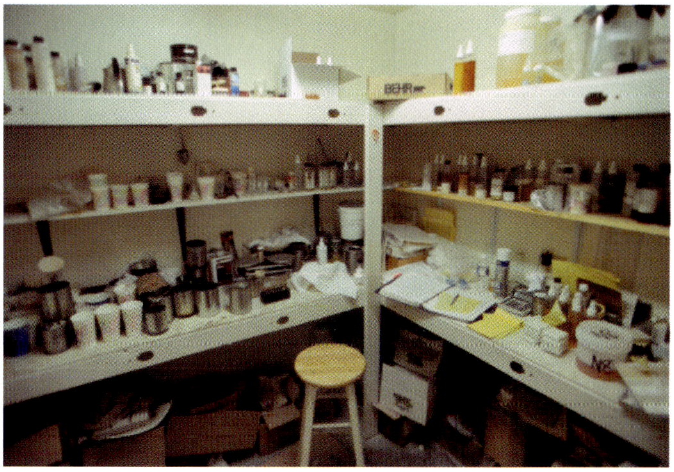

The base material is heated and mixed in monster vats (stainless steel Navy food processors).

After the scent, the color—FDA-approved for skin contact.

Out of the vat, into the mold.

The hot wax dribbles into individual molds.

Like little loaves of bread, the wax cools on racks overnight.

Neatly boxed, the product is ready for retail.

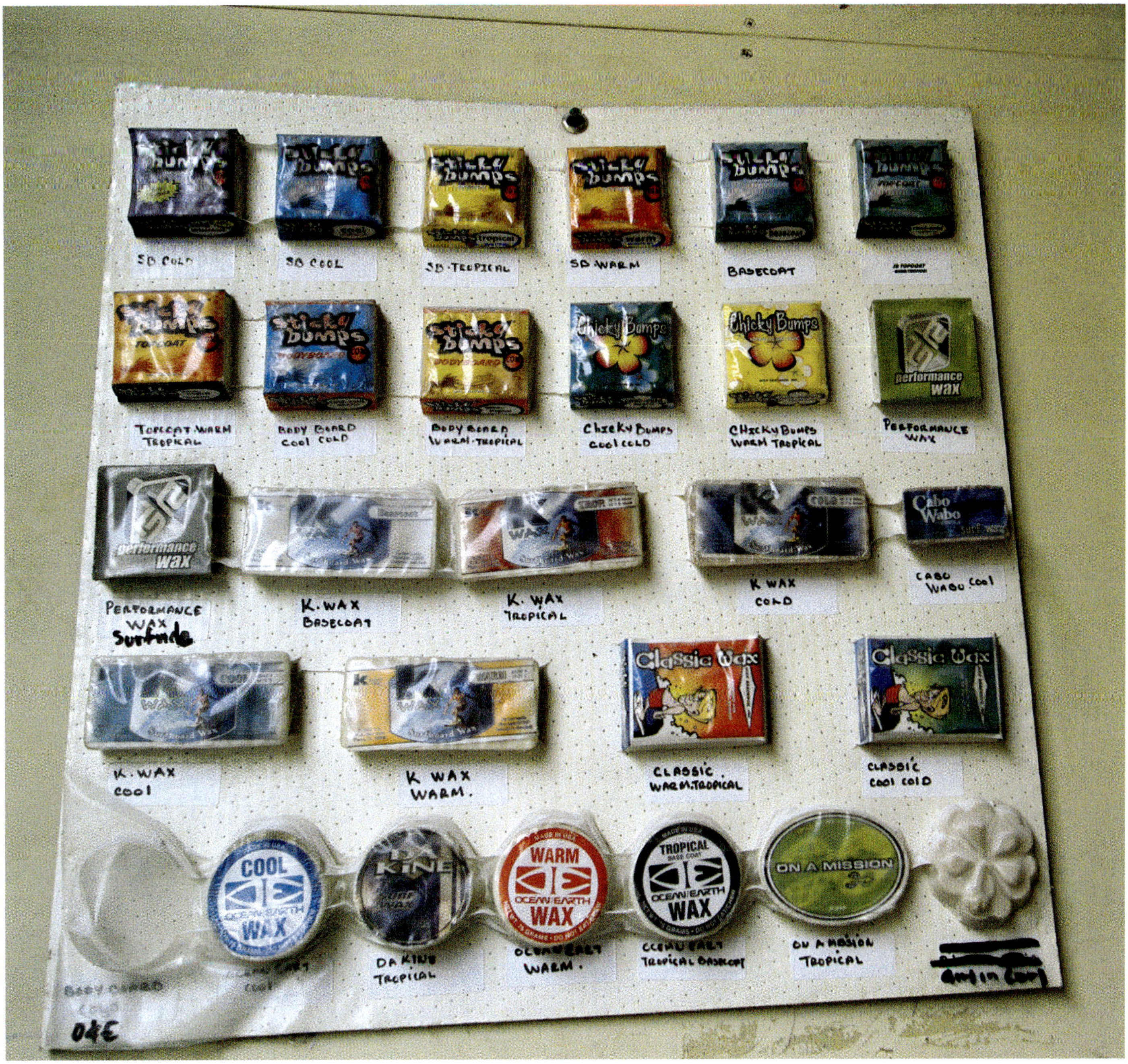

Various Sticky Bumps products on the wall at the factory.

Now in the twenty-first century, the Big Three once again don't seem to think that marketing their products in *Surfer* magazine is altogether necessary. Wax ads are almost non-existent in The Bible these days, as apparently there is no room for them among the ads for World Wide Wrestling, Master Craft tow boats, Speed Stick anti-perspirant, and VW that make *Surfer* look more and more like *Details* with every passing month.

But just because there are no advertising smoke signals doesn't mean the market isn't on fire. There are some pretty bizarre figures floating around about the size of the surf industry. Reef Brazil's claim that there are 17,000,000 surfers in the world was taken as gospel by anyone who surfs Trestles, but discounted by the rest of the world.

Randy French at SurfTech swears by a number of 400,000 for the total yearly surfboard market and if that is true, then what is the wax factor? How many bars of wax are applied to how many of those surfboards, how many times a year?

"Wax Research is blowing up as we speak," John Dahl said in July of 2003. "Customer sales were conservative during the Iraqi situation this spring. During summer our production has been completely swamped. We ramped up to 20,000 bars a day and are still trying to catch up. Our transition from renegade, backyard manufacturer to legitimate business has occurred over the last twenty years. To be a manufacturer in the state of California (especially surf wax) is insane. Workmen's comp, product liability, worker's benefits, vacations, bonuses, and retirement are challenging but when you can be in the water in five minutes it's a wonderful insanity."

Zog's has also made the transition from backyard brews and tuna cans to a thoroughly modern manufacturer.

"Our current facility is approximately 6,000 square feet at ground level, with an additional 2,500 square feet of mezzanine for storage," Zog said. "We have been here since 1983. We are currently operating with only eight employees, but our operation is very efficient. In addition to wages, the employees are provided with a company-paid health and dental plan. Furthermore, employees who are in their third year also receive Sep Ira Retirement benefits that are contributed by the company to whatever account the employee specifies. In the fourth year, additional bonus money based upon the profit of company is also distributed to employees."

Regarding production, Zog's is not quite at the level as Wax Research but they are putting out a lot of wax.

"Our maximum daily capability is 16,000 bars a day for a 10-hour day," Zog said. "But we are currently producing up to 12,000 bars per day at this time. With ample storage for finished product we try to maintain a relatively constant production schedule year 'round. Our finished goods inventory for Sex Wax will become excessively large during the fall and winter months, but will drop extremely low during the summer. I am not sure how much wax John is producing annually, but 20,000 bars a day is a lot of production. For the entire year of 2002 we sold in excess of 2,500,000 bars of wax. We do not work on weekends, and with holidays and down time, this represents an average daily production rate of about 11,000 bars per day."

When asked these same questions about production and sales at Mrs. Palmers and where it all goes, Alistair Cook climbed to the top of a 50-foot high pallet of stacked surf wax boxes and threatened to jump. When he calmed down, Mr. Cook held court on the trials and tribulations of making big money selling something for pennies.

"We do not really want to talk about volume per day," Cook said, "as I don't really feel this is a fair question to ask us in a competitive surf industry where Quiksilver, Billabong and Rip Curl wants to put their name on anything they possibly can to maximize their shareholder profits. Wax is a business of large volumes and small dollar terms, as each unit sells at a very low price. At Palmer's we have entered into the clothing and footwear business in recent years, and the bottom line is one 40-foot container of footwear holds about 6,000 pairs of shoes and is worth approximately $400,000 to $500,000 wholesale to the surf shops. To sell the same dollar term as 6,000 shoes you need to sell a million bars of wax. Wax is a specialized field and it cannot be made in China, so the margins are also nowhere near as good as the apparel industry.

"The history of the wax business shows hundreds of people have tried to put their names to wax, but the bottom line is you need to be doing millions of bars a year just to make it viable. It takes years of experimentation and development of specialized equipment to make it. It is also a labor-intensive and messy business, which has at this stage kept it a bit of a specialized business you only get involved in for the love of the sport and lifestyle.

"I think the key thing that probably Zog, John Dahl and ourselves have in common," Cook said, "is we all deal with wax in and out of our factories by the ton and container-load. It comes in bulk by forklift and leaves in a smaller bar but also by forklift on pallets and containers. We often find ourselves noting it comes in by the ton, goes out by the ton but it never comes back.

"We could all answer your question of how many we do but no one can answer the question: 'Where does it all go?' Rubbish bins, old boards, carparks, deep into carseats, back into the sand, ocean and earth (where these materials initially came from)—no one really knows. At Palmers the key thing we have learned after fifteen years of hard work is someone must be getting a hell of a lot of good surfs and great waves in many different parts of the globe. Unfortunately for the three wax makers, we are helping to make these surfs more enjoyable for most surfers but we probably aren't getting our fair share, as we have been too busy dealing with making it and running our businesses. Your Gordon Merchants, Bob Hurleys,

Greenie, Claws, Warbricks, etc., all entered the surfing industry at the same time as Zog and John Dahl so I can't help thinking some people decided to make the right products and others decided just to make wax."

Jefferson Wagner Tours Waxlab

John Woodward Shares (some of) His Recipe

During my travels and research efforts, I came across John Woodward, owner and founder of Waxlab. Waxlab is a private label wax manufacturer in Canoga Park, California. John has roughly twenty wax clients, each with its own mold, compound, and scent. Waxlab is situated in a light manufacturing district in a concrete tilt-up building.

All the commercial building fronts appear the same, but opening the front door of Waxlab is a true experience. I made the mistake of wearing my decent shoes. My first step onto the sticky mixing floor pulled the right shoe off. After a quick adjustment and the old, "no problem, John," I was on the tour. John is producing 70,000 to 80,000 bars of wax each month. The factory has several sections separated by 2" x 4" framing and drywall covered with surfing posters. Taking a deep breath rewards one with the fragrances of coconut, banana, and bubble gum all in one.

During the tour, John told me the story of his initial effort making wax in 1985. His first wax, Ganja, came in a baggie, was square, and had a Thousand Oaks, Calif., hometown on the label. A later label had Zuma Beach, Calif., on it. After twenty years and five moves in and around the Los Angeles area, Waxlab has landed and is literally glued to the concrete foundation.

John was nice enough to share with me the secrets of the different wax compounds. I promised I wouldn't reveal the names of all the private labels, but they are all pictured somewhere in this book. He did mention that the same truck delivers some of the same compounds to Zog's factory up the 101 freeway, an hour- and- a- half north.

After the tour, John explained in detail his own wax manufacturing process. I assume it is somewhat similar for the Big Three—Wax Research, Zog, and Palmers—but on a larger scale for them.

John begins by heating one compound (the contents of which are secret) to 300 degrees in a 55-gallon drum. The melted mixture looks like beige lava from a Disneyland ride. Then an 11-pound slab of straight paraffin is added to the superheated compound.

The mixing of this slurry is done mechanically with a Milwaukee ¾ hp drill with a paddle attachment at low to medium speed. Scent and colors are added as the compound cools so the scent is not lost due to the heat.

The final mixed compound is ready to pour first into pitchers and then into individual wax bar molds. There are a lot of different shapes for such a simple product. Each brand has its own shape and wrapper. The hard molds are made out of aluminum on a CNC (computer numerically controlled) machine. The hard molds are then used to make the PVC (polyvinyl chloride) final mold that the liquid wax is poured into for cooling. Once in the mold, it is off to 32 degrees Fahrenheit (overnight in the freezer).

The cool brittle wax pops out of the plastic mold easily with just a twist of the plastic tray, usually four to eight bars per tray, depending on shape and complexity. The product goes to a clean part of the factory where it is boxed and set on shipping pallets, then off to the distributors.

With the necessary care and patience, the result will be a quality product. Wax manufacturing is not really a difficult process, but it is really hot and messy.

Chapter Nine
Three Collectors and Their Stories

This book would not be complete without mention of some wax collections from different surfing areas. During the three-year research phase, I discovered some rather serious collectors. In order for me to call someone a collector, they needed to have one hundred or more bars of a different nature or manufacturing style that did not include private label waxes produced by one of the Big Three companies (as there may have been hundreds of private label waxes produced by the Big Three).

Three notable collectors in the western United States were of particular help to me—Andrew Tamm in Hawaii, Casey Mills in Oregon, and Kevin Hall in California. Without their efforts and help, the extensive picture content of this book would not have been possible. Each collector has a few rare bars that neither of the others has.

This chapter is dedicated to the collectors and their rare pieces. The photographs were taken by the collectors themselves and may not reveal complete details. Each collector has provided a short biography.

Kevin Hall
We'll surf forever

Ahh...the smell of surf wax. My name is Kevin Hall, and that smell is one of my youth. I live in beautiful Southern California, near the coast, and attended Newbury Park High School in the 1970s.

In high school back then, you were either a jock, a stoner, or a surfer. You had to choose, and I chose surfing. We would spend every possible minute at the beach surfing and hanging out. It was my home away from home. We would hang at the local surf shops when the surf was flat and just talk surf.

The surfboards were progressive and diverse, and so were the surf shops. Wax always caught my eye, with its vibrant colors and that smell...

Around the 1980s, I started collecting surfboards. I would find them everywhere. In rafters, at garage sales, from friends. Longboards or shortboards—I wanted to try them all. Sometimes I would score some old surf wax, too. This really intrigued me.

My surf wax collection started just by curiosity to see how many sizes, shapes, and colors I could find, and the boxes started to fill. No matter where I am in the world, I will check every time I see a surf shop. To this day, I am still surfing and still adding to my collection.

I plan to keep this collection as a small piece of our unique surf history. The following pictures show some of my rare and exotic waxes.

My wife, Robyn, and I, my daughters Laurel and Megan, and my son Cameron still reside in Southern California. We plan to stay here and surf forever.

—*Kevin*

Balance tub

Bali Bagus

Bart Simpson

Baliga Paraffin

California Surf Museum

Climax

Da Best – mini

Easy Rider

Esso tub

X-tremes

Free surf

Gravity Control

Gravity Control

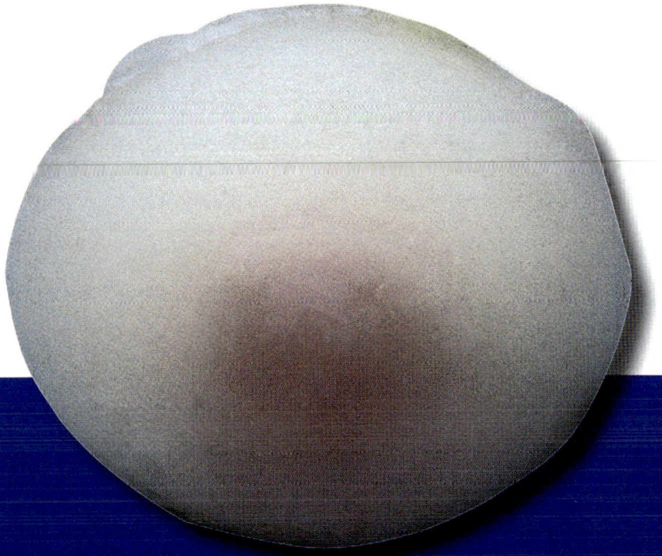

Handful

K Wax

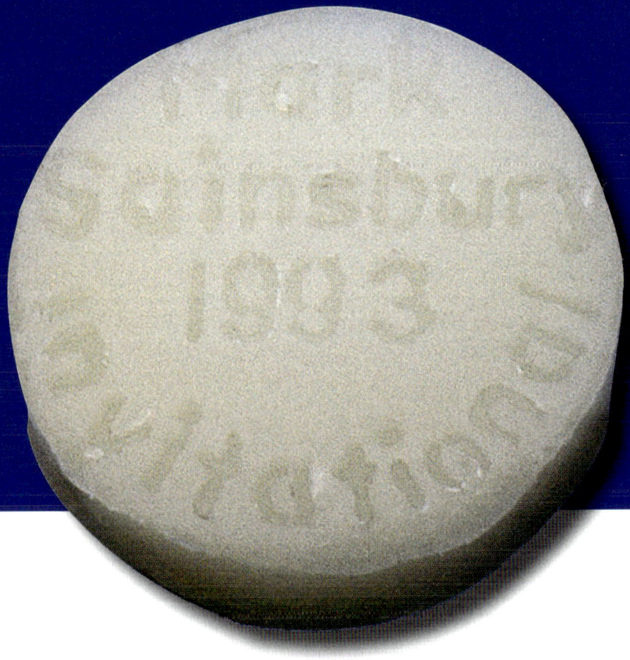

Mark Sainsbury

Max Wax

McDonalds

Nat's Wax

One Ball Jay Skull

Original Green Stuff

OZ Sea

Quicksand

Ridgy Didge

Sticky Feet

Stik Wax

Superwax

Surfers Choice

Survival

Swix

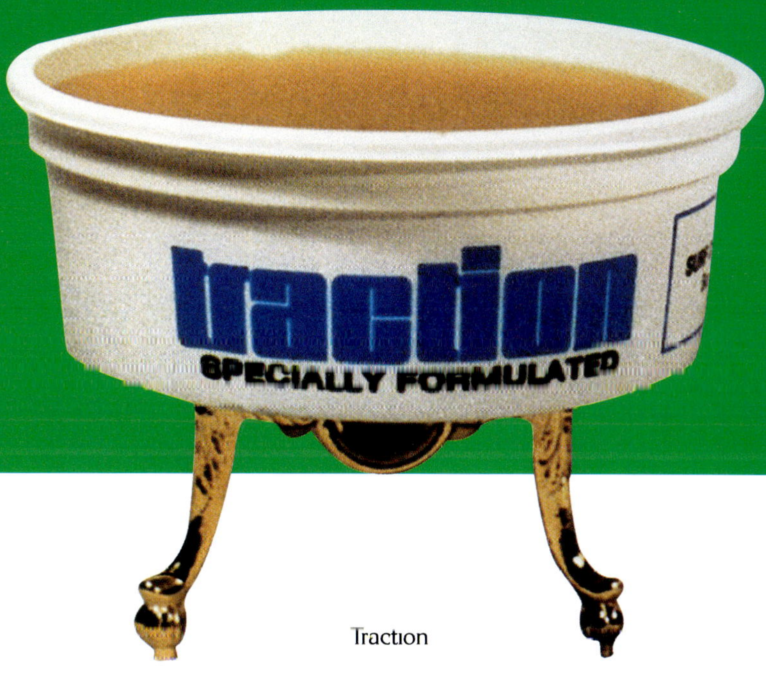

Traction

Triple J

Troff Lolly

USO tub

WIZSTIX
100g
SURF WAX
100g

Wizstix

Z's
SURFWAX
"NO Wimps"
COOL
Do Not Eat or Chew
Z's Surfwax
P.O. Box 1367
Oceanside, CA 92051

Z's

Casey Mills

Definitely not for sale

My name is Casey Mills and I grew up "in the Valley," down in Southern California. I currently live in Keizer, Oregon, with my family. My first experiences in the ocean were from Zuma and Malibu to Huntington and Newport. I started out body-boarding at age five and as I got older, the body-boarding grew into surfing and skimming. I currently do all three with my children.

I started collecting surf wax and surf stickers at age ten because they were way cheap, smelled great, and had awesome colors and shapes. When I got a bit older, we covered my dad's Toyota 4-Runner with hundreds of surf stickers. You probably saw us from time to time.

See, I could take two bucks into Zuma Jay's, Val Surf, or Jack's and come out with a ton of wax—what a deal! I have over 500 bars of wax and about 150 different kinds—200-plus if you count different versions of the same brand. The collection went from LA to Kauai (lived through Hurricane Iniki in 1991) and is now with me here in Oregon.

Although I have been offered a good chunk of change for what I've got, I have never thought about selling it. This is part of surfing history, and that is quite nice; I want the collection to remain in the public eye for all to enjoy. Feel free to visit for a glimpse, but know that history is not for sale. We have, however, been known to trade!

—A Hui Hou. Mahalo Nui Loa,
Casey

Buzz Wax

Cabo Wabo

Killer Bees

Murph's – three different varieties

Wahine Wax

Wax-man

Andrew Tamm

Finds rare waxes around the world

I was born and raised in Oahu, Hawaii, with roots going back in the islands five generations to the island of Kauai. Surfing was and is my passion.

I was first introduced to surfing at the age of eleven by my older brothers, who decided to take me out surfing one day after months of constant begging. Since that fateful day, surfing has become my passion, which has kept me in the water on a weekly basis for the past three decades.

Starting off in the 1970s with hand-me-down old surfboards, I learned to cherish old things. At that time, most surfers used any type of wax they could get their hands on, from candle wax to Parowax and, on a very rare and special occasion, a bar of good old Sex Wax or Lightning Bolt Wax.

In the '70s, the local boys preferred Sex Wax with the brown label and the orange-colored wax. Cruising the beach between surf sessions with friends, I got in the habit of trying different surf waxes that came out, and enjoying the various fragrances of each bar.

The everlasting desire to collect wax really started when I went to college in San Diego. At the Star Surf Shop in Pacific Beach, which was owned by a Hawaiian guy who moved to the mainland, I was amazed to find all types of surf wax brands. California had a maze of surf shops as compared to Hawaii, and it was great fun to check out the waxes out in surf shops.

Wax was something I could afford. Sometimes I only had a couple of dollars, but it was enough to get a bar of wax.

From this point, I started collecting as many different brands and colors of surf wax as possible. Later, when traveling around the world in search of the best surf spots, I would collect bars of wax that were missing from my collection. I found some of my most prized pieces in shops in Tahiti, Indonesia, Mexico, Japan, and Fiji.

I have always felt that the most essential equipment for surfing waves consisted simply of the surfboard and the surf wax. Without either of them, we would not be able to stand up to surf. It is because of my passion and love of surfing that I have become an avid collector of vintage surfboards and surf wax. I stick simply to the basic two elements needed to get me going surfing…surfboards and wax.

Sharing my collection with friends and surfers is something I really enjoy. It brings back memories and stories, and most people are surprised to learn that there are so many different shapes, brands, colors, and smells of surf wax made in the last four or five decades.

I hope you enjoy the pictures of some of my rare waxes.

—Malaho,
Andrew

Budweiser – large size

Frog

Friends of Da Hui

G - Land

Gerry Lopez

HLF

Harry Hinilulu

Hasa

Hawaiian Poi

Hawaiian Poi

Heritage

Ho'okupu

Hot Stix – original

Medicated

Orange Stuff

Source

Sticky Lips – round

Surf Wax – in a bag

Tiare girl

Triple Crown

TRS
SURF · CLUB · HAWAII
Taki & Rex's
WAIKIKI, KAHAKAI

TRS

VooDoo
SURF WAX
VooDoo
SURF WAX

Voodoo

Chapter Ten

The Master Collection - 250 Wonderful Old Waxes

Notes On the Master Collection

By Ben Marcus

Jefferson Wagner has collected surf wax since the mid-1960s and has owned a surf shop in Malibu since the mid-70s. He has sold countless bars of wax, at the rate of about two cases a week, and has kept at least one sample of every brand sold on his shelves. His collection includes old and rare bars and consists of the actual waxes in their original packaging.

"That is why this collection is so unique," Wagner said. "It isn't just the label. This is the entire package as it was presented for resale."

When Wagner started surfing, household Parowax was the substance used on boards for traction. There were no products known as surfboard wax.

"It was in boxes with three leafs," Wagner said, contemplating his prizes. "We would cut it up into any kind of form you could imagine. And we would split the cost of a gallon of gas when we got out there."

Wagner's collection includes Parowax that was cut into cookie-cutter shapes—stars, clovers, lips and such.

"That truly is surf wax. Halloween wax lips are hollow. This is the real stuff."

Wagner purchased some of the older specimens at Val Surf in North Hollywood, as there were no surf shops in Malibu then. Some of his waxes are of unknown origin.

"I surely don't know where that came from," he said of one bar. "In a few years someone can carbon date it. Remember, wax is all dead dinosaurs."

The collection documents the growth of environmental awareness in surf wax packaging, from plastic bags to recyclable, biodegradable paper and cardboard.

Wagner's rarest bars are an original John Dahl product and a Lightning Bolt Wax. Long Ryder Surf Wax is another unusual bar in the collection.

For a while he sold his own brand, Zuma Jay Surfboard Wax.

"That was originally done by [the late] Con Colburn in the '70s. He had the molds and the processing plant down there in Venice. He private-labeled for me and for the other '70s surf shops in this area and I paid him directly. Con was a good guy and he was really into all this.

"We probably sold a thousand bars a year—not a lot back then. There weren't as many people surfing. And it kept Con happy. Kept him going on his wax operation."

Randy Dowler also sold waxes to Wagner, including Waxx On, Kangaroo Goo, Koala Bar, Breakaway, Piranha, Super Sticky, and Styx.

"He would come in here during his sales trips up and down the coast and kind of get a feel for how it would sell and what kind of response he would get from customers because I was always around when he came in. He would always ask me, 'What do you think about this?' and 'What do you think about that?' We had a lot of fun with all his different wax labels. He was in and out with a lot of these different waxes that I have."

"Each one of these waxes," Wagner says, "is somebody's dream that they made happen, or they let go of."

Jefferson Shares His Lifetime Collection

The following pages contain what I believe in 2006 to be the most extensive collection of surf wax in the world. No doubt a few readers of this book will be compelled to write me and express concern or criticism that some bars were left out or not mentioned.

Collecting is a personal thing, and these are the surf waxes that resonate with me. If every existing wax (past or present) were included, this book would be too big to pick up. Each wax brand eventually could have four to five variations based on temperature, and some labels have only one production run. I therefore selected the waxes that are the most important in an historical context and that had a retail shelf life somewhere.

To those who collect, I say, "keep on it." For those who are starting, I advise, "Keep them in a cool dry place inside Baggies."

Nothing is more satisfying than hearing a mom or dad come into my shop, look at the collection and say, "Hey! I used to use that stuff!" I can tell from their voice whether they got a great wave or got pummeled that day in their memory. They pass to their kids the same sense of excitement.

That's who this book is for.

—Thank you
Jefferson "Zuma Jay" Wagner

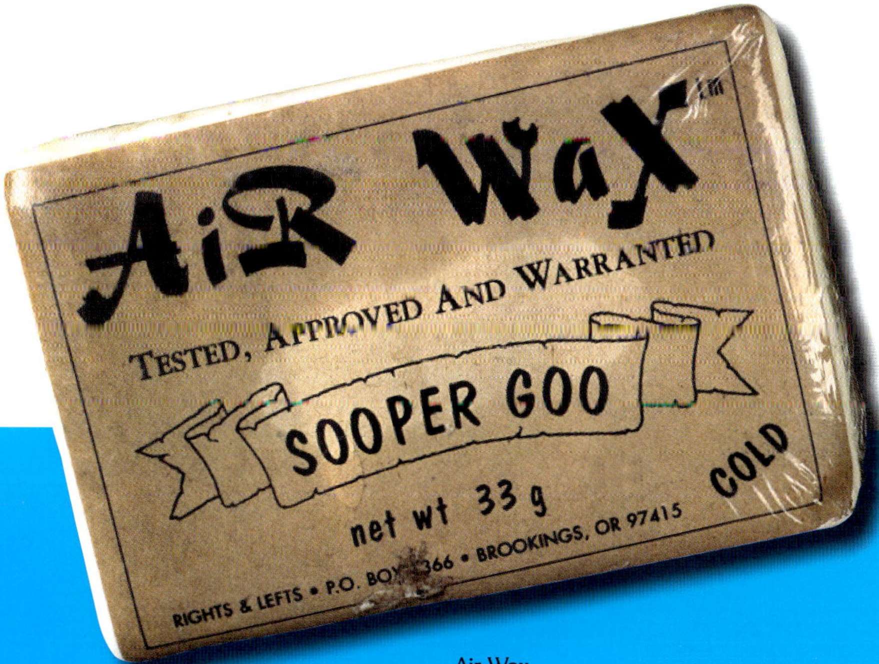

Air Wax

Astro

Bad Wax

Bali Brew – original

Bali Brew – large size

Original
BANANA
SURF-WAX
831 WILLIAMSTON ST. SUITE 175 VISTA, CA 92084
BANANA BROS.
NOT A FOOD PRODUCT
WARNING · DO NOT EAT OR CHEW
TM
wax with a peel
COOL WATER
65 GRAMS

Banana

BANG!
LITE
WARM WATER

Bang!

Bearfoot

Big Block

Big Pecker's – competition

Big Pecker's – square

Black Market

Blue Link

Bodyline Hardware

Bolt

Break-away

Bubble Gum – original, animal, beeswax, warm, and new

Bubble Gum – green label

Bubble Gum – round, sticky

Miller, MGD, Foster's, Bud, Watergirl

California Summer – large

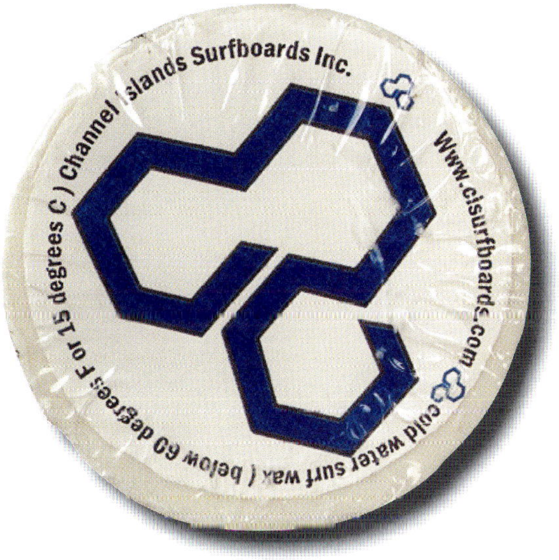

Channel Islands

Chicky Bumps

CHP

Classic – original

Classic – new

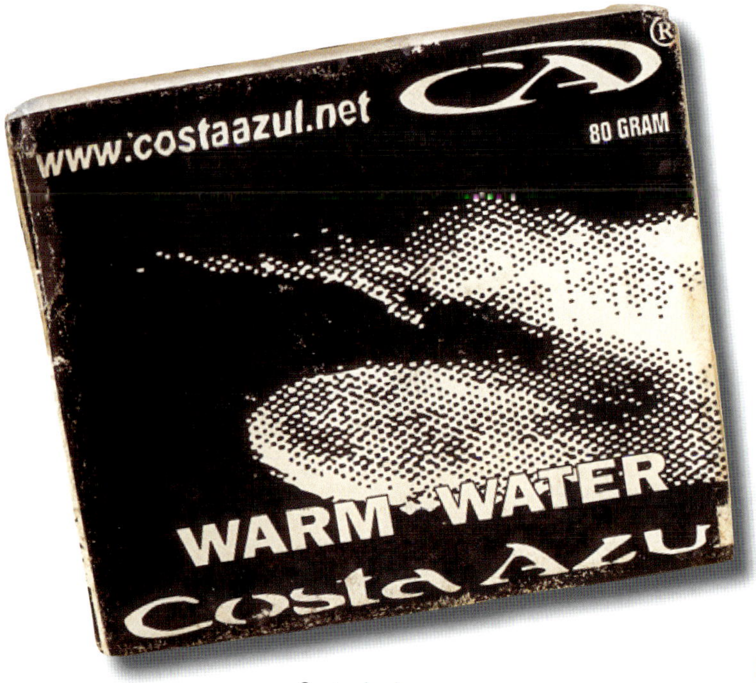

Costa Azul

Control Wax – square

Control Wax – round

Quiksilver, Split, Lava, Coors Light, Hawaiian, Val Surf, Channel Islands

Da Kine

Da Right Style – Beez Wax

Dr. Ball's

Drugs – topcoat

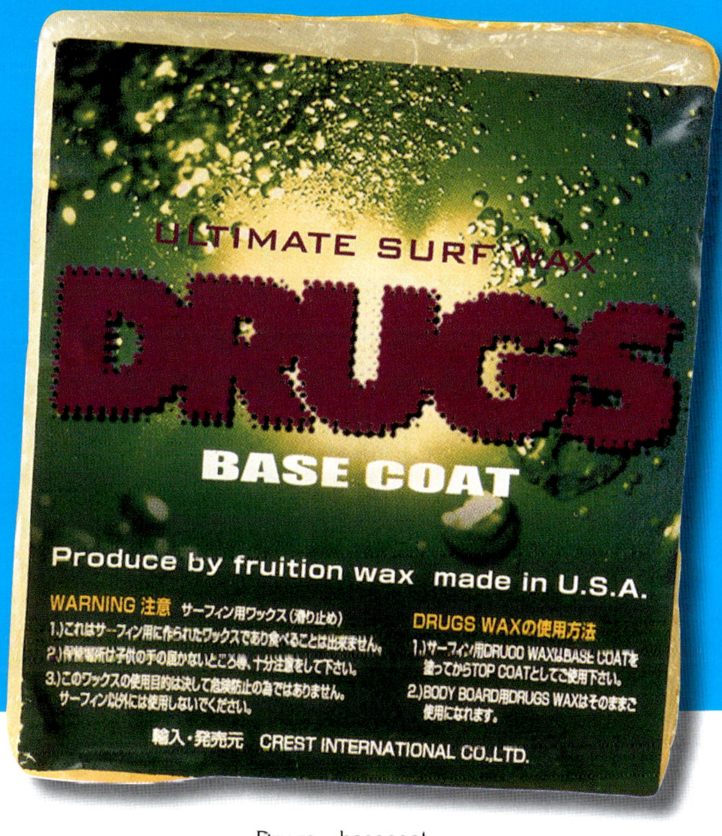

Drugs – basecoat

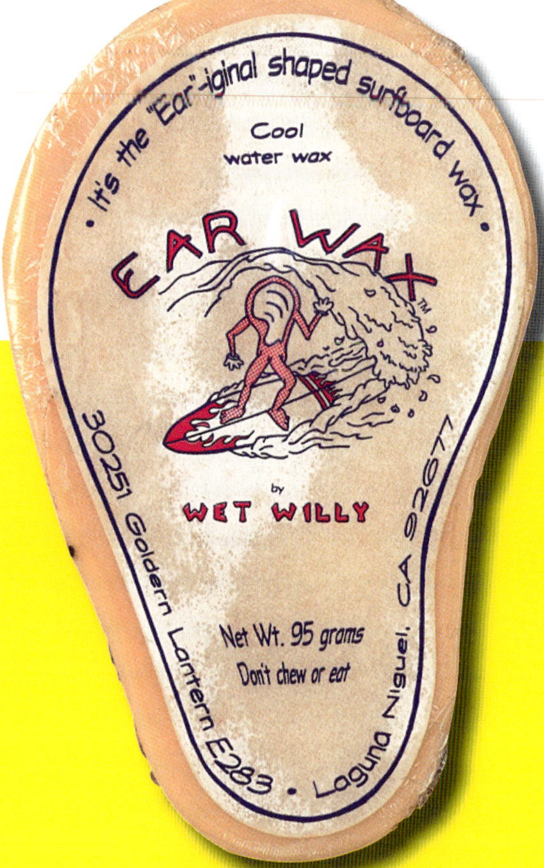

Ear Wax (front)

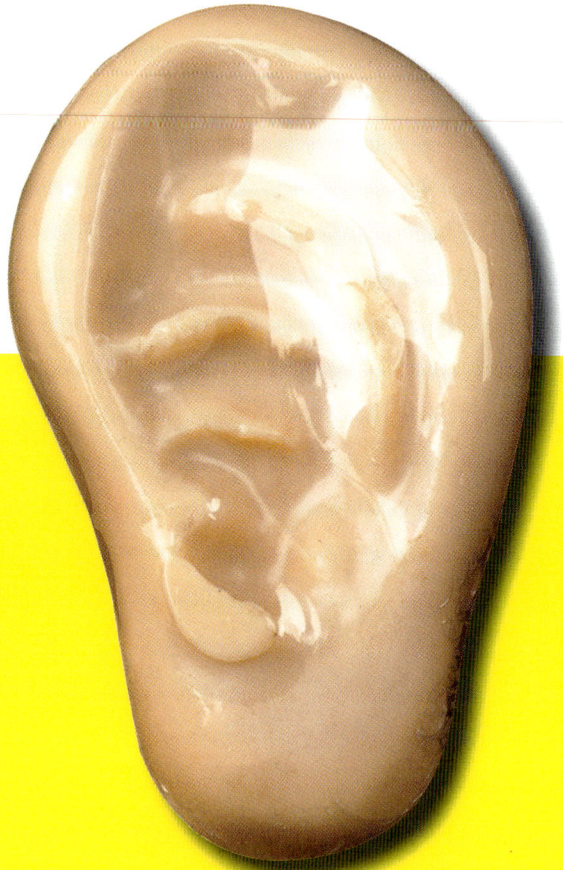

Ear Wax (back)

Elmer's

Extremely Sticky

Famous

Five Daughters – paper

Five Daughters – box

Five Daughters – sticky

Foot Suckers

For Perfection

For The Grip

Fruition 999 – original

Fruition – new

Fubar

Ganja – Malibu, Thousand Oaks, Sappy Bead

Ganja – black label

Global Lip Grip

Global Soul

Good Stuff – original

Good Stuff – new

Grade "A" Skim Milk

Grippers

G & S Original

G&S – large

Hot Stix

Hawaiian Hot Stix – original

Hawaiian Hot Stix – 1988, 1989

Hawaiian Lava – molten formula

Hawaiian Lava – 1995 registered

Hawaiian Lava – surf wax

Hawaiian Style – poi

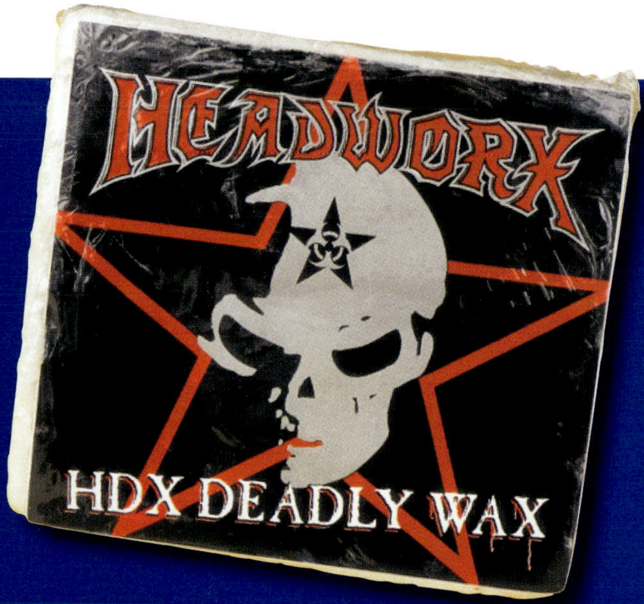

Headworx

Honey Wax – warm, cool, cold

Hooters (front)

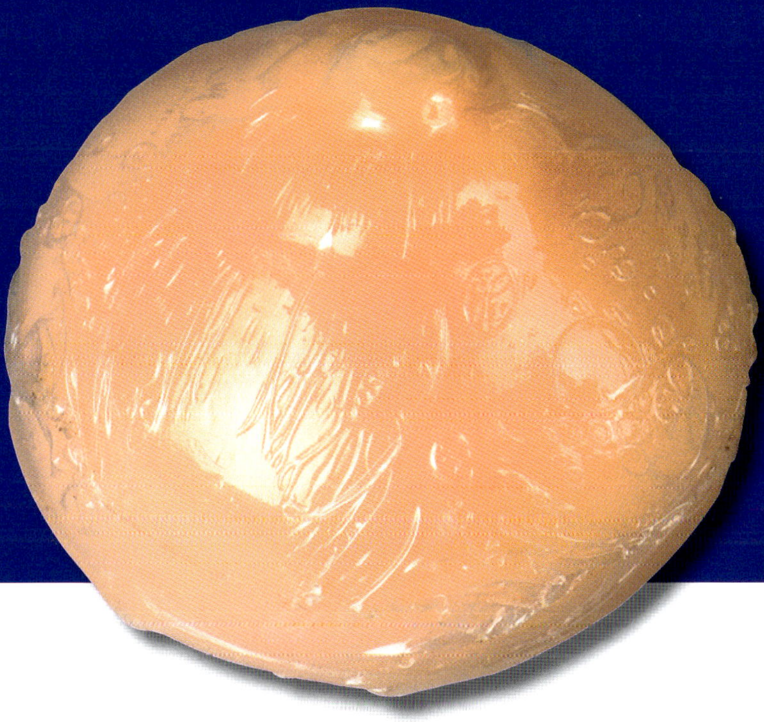

Hooters (back)

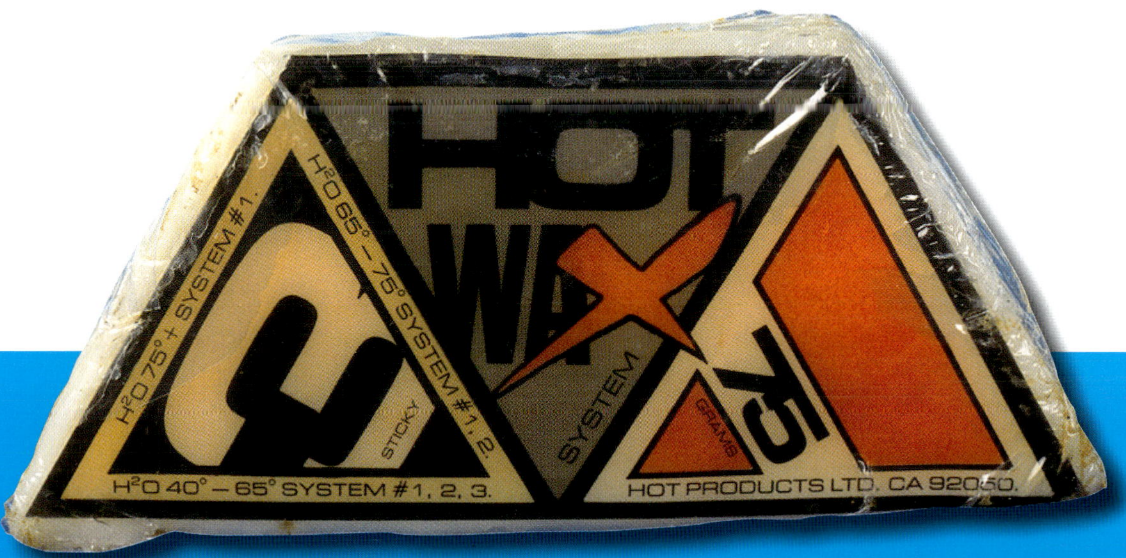

Hot Wax

Hobie – team

Humpinwax

Hula Deck

Hydrolight – square

Hydrolight – round

Icky Sticky Ooey Gooey

Intersurf

Just Wax!

K Wax – round

Kangaroo Goo

Kangaroo Goo – king size

Killer Bees – bag

Killer Bees – paper

Killer Bees – bloody stinger

Killer Bees – premium

Killer Bees – original

Kin – plastic applicator

Koala Bar

Len's

Local Motion

Longboard – original

Longboard – new

Long Ryders

Lorelei's

Mad Wax – original

Mad Wax – new

Magma

Matuna's

Max Wax

Mighty Mounds

Mike Purpus

Moon Cresent – Japan

Mother's Milk (front)

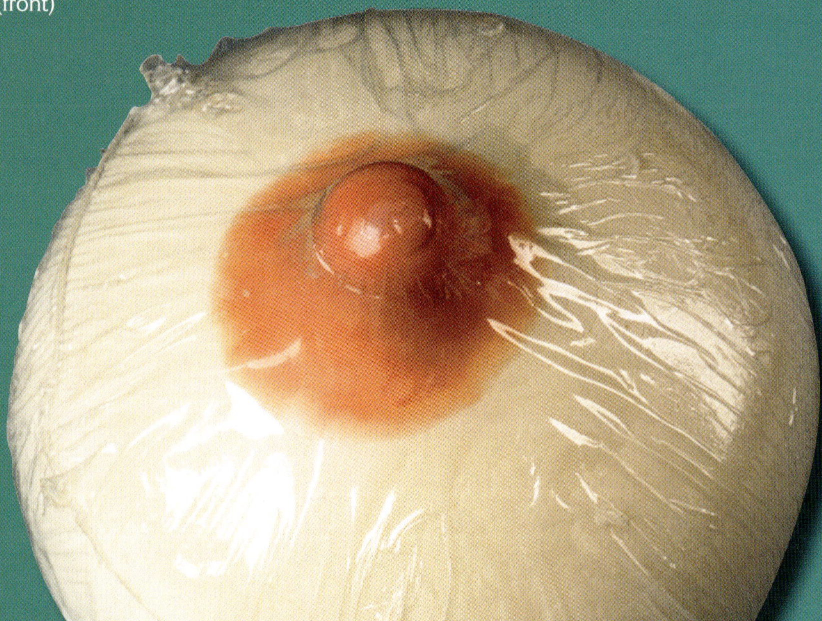

Mother's Milk (back)

Mrs. Palmers – plastic wrap

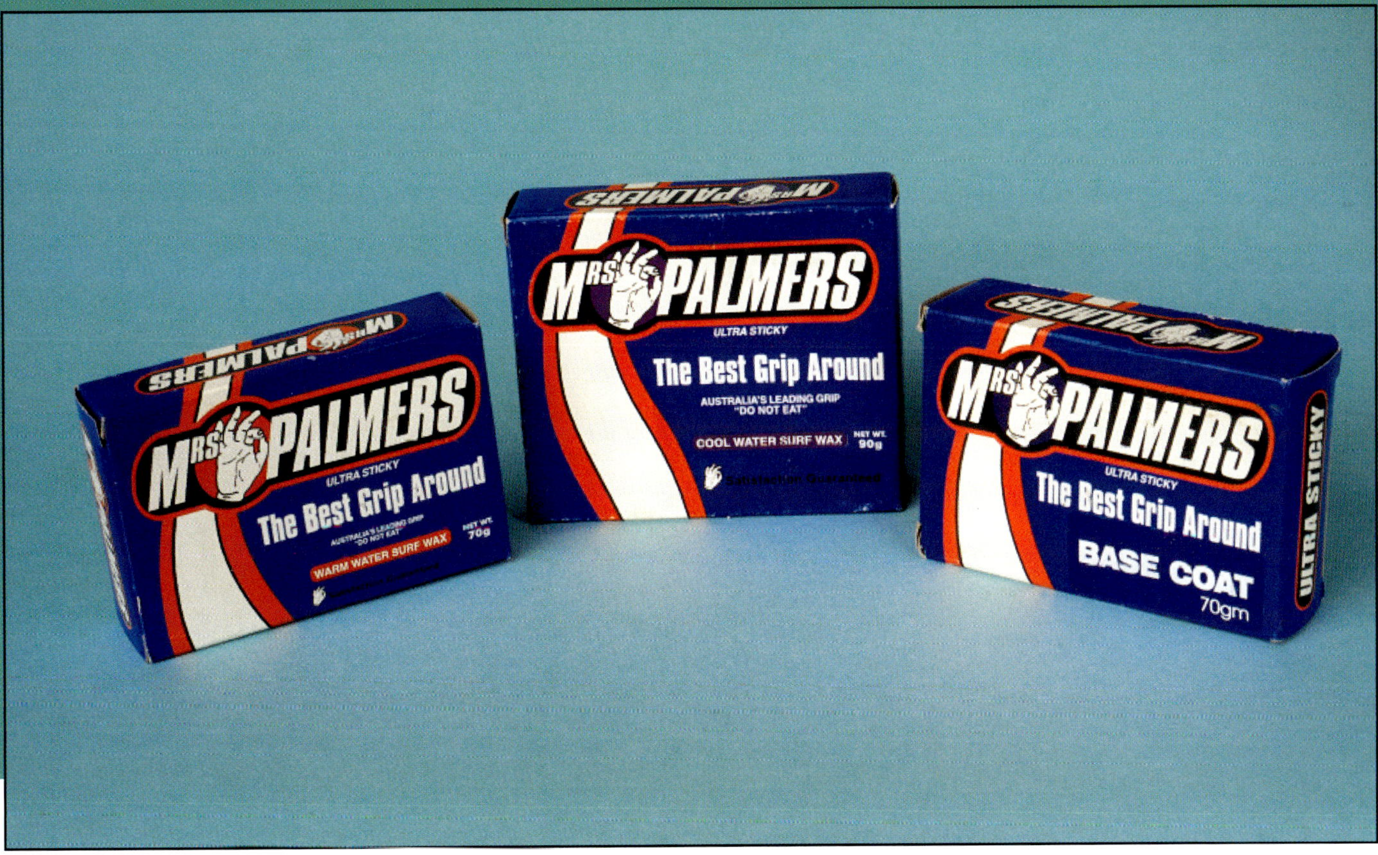

Mrs. Palmers– new, large, and basecoat

Murph's – original plastic applicator

Murph's – new plastic applicator

Natural Progression

Nat Young small and large

No ka oi

On A Mission – basecoat

On A Mission – original

On A Mission – green

Ocean Pacific – original

Ocean Pacific – new

One Ball Jay

Four decades of Parowax

Performance Wax

Piranha

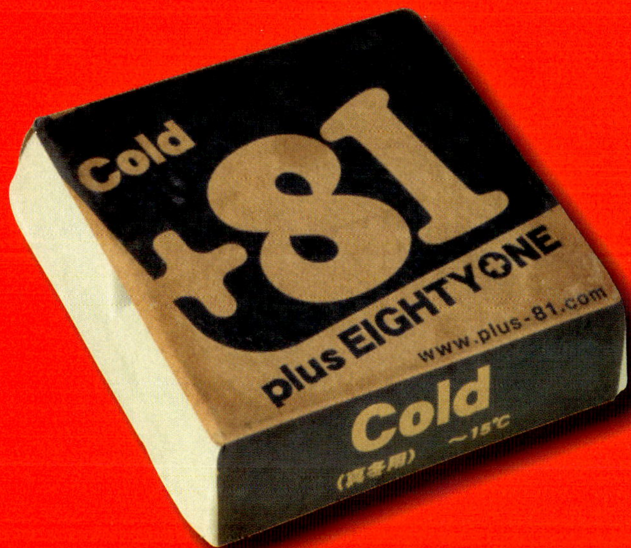

Plus 81

Poon-tang's

Poorboy – tub

Poorboy – round

Poorboy – bar

Powderkeg

Power V8

Progrip

Pro-lite Classic

Pro-lite – cold

Progressive Expression

Pure Girl

Quiksilver – round

Quiksilver – square

Quickie Bumps

Ransom

Rat

Really Tacky

Rebel Soldier

Rebel World

Red Palms

Red X

Rocket – plastic applicator

Rocket – refill

Roxy

Sack-O-Shit (front)

Sack-O-Shit (back)

Santa Cruz

Shark Bait

Shark Bait – basecoat

Shark Repellant

Sick Pig

Shred

Slyme

Slipcheck – can

Smack

Smack – basecoat

Southern Comfort

Star – parowax cut-out

Sticky Bumps

Stick-it – plastic applicator

Sticky Bumps – epoxy boards

Sticky Bumps – classic

Sticky Bumps

Sticky Bumps – tropical

Sticky Fingers

Sticky Gum

Sticky Wax

Storm Rider

Sugar Bear

Super Sticky – tub

Super Sticky Bumps

Surf Dope

Surf Line Hawaii

Surf Research

Surf Wax

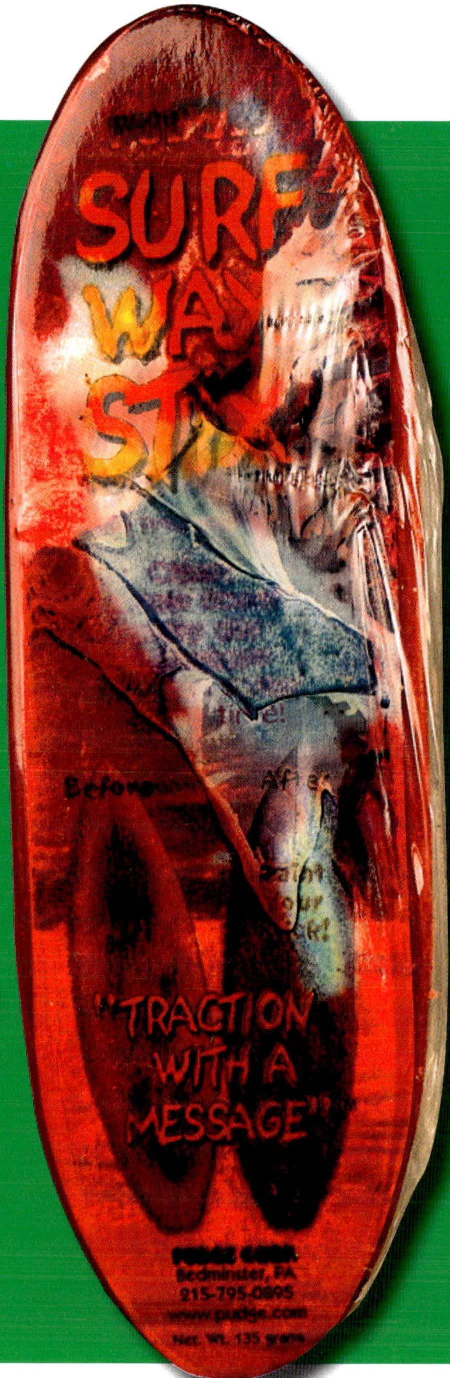

Surf Wax Stix

Surfari

Surfers Alliance – new

Surfers Alliance

Surfers Choice – original

Surfer's Choice

Survival

Sticky Lips – new (left) and original

Surf Wax Museum

Takwax

Terra Wax

Tiki Wax (back)

Tiki Wax (front)

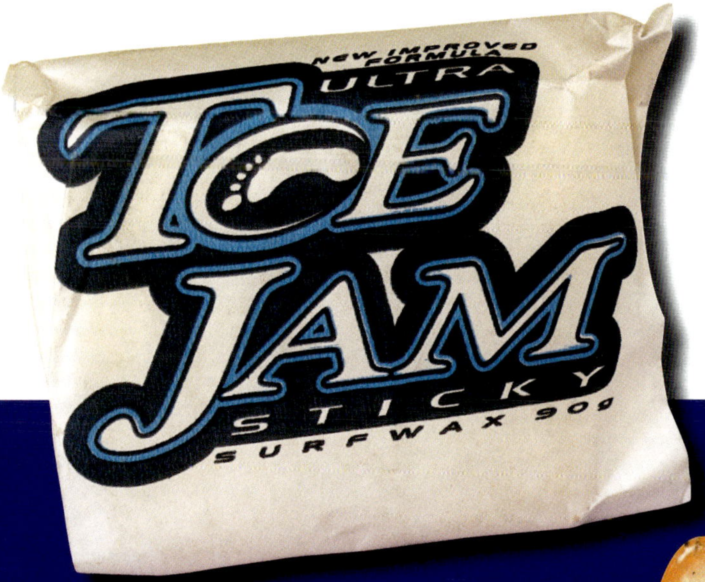

Toe Jam

Toe Jam – original

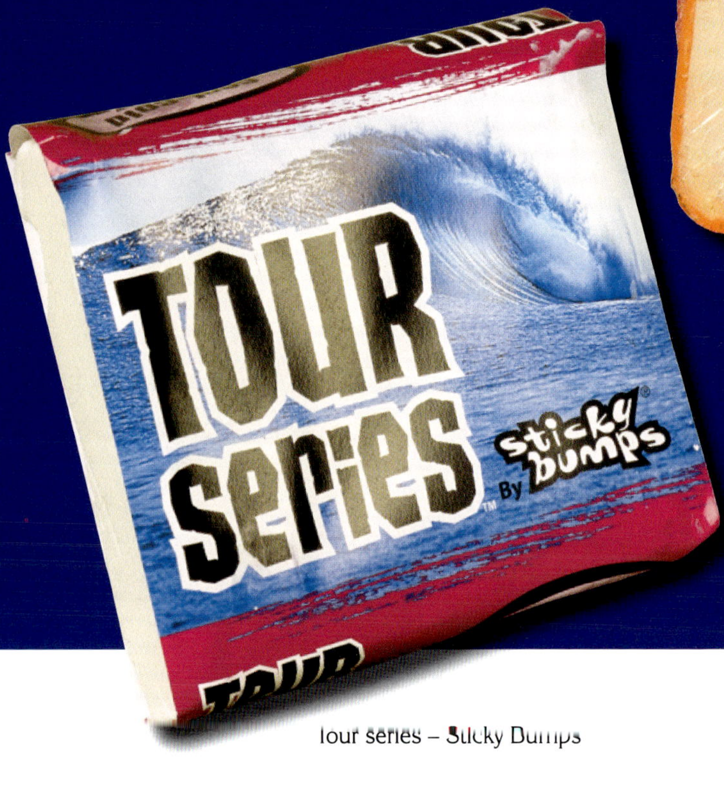

Tour series – Sticky Bumps

The Big Bar – tub

Tropix

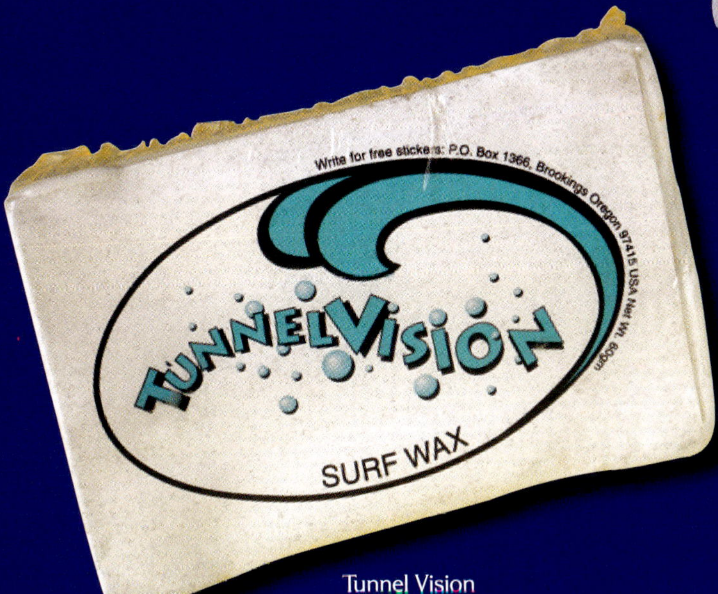

Tunnel Vision

Tyler – original and new

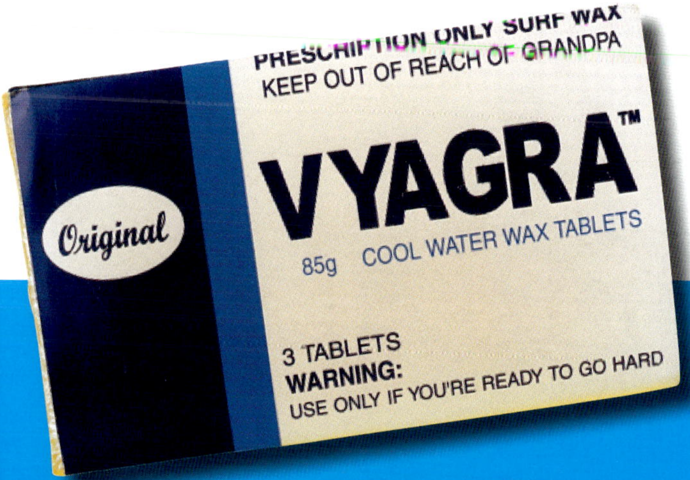

Vyagra – original

Vyagra – nubs

V8

Wahine

Wax Bomb

Wax Connection

Waxmate

Wax Pak – plastic applicator

Wax Research

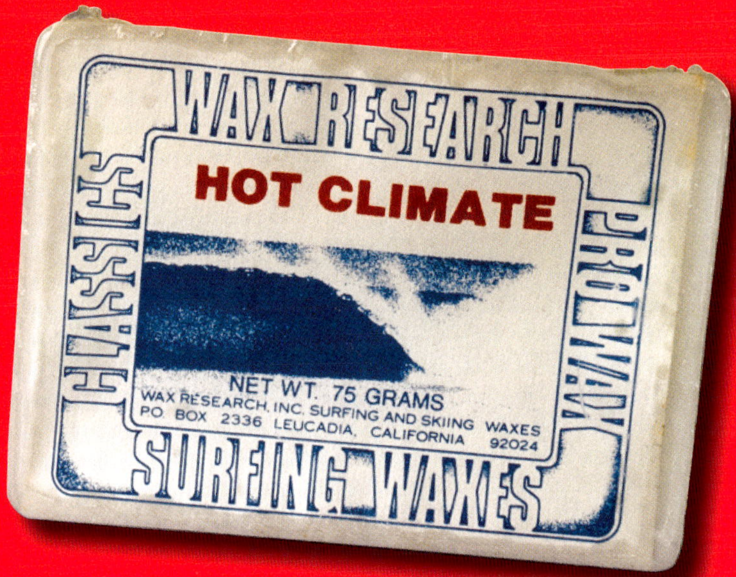

Wax Research

Wax Research

Wax Research

Wax Research basecoat

Waxx-on

Waxx-on

Waxx-on – basecoat

W.E.S.T.Y.

Wet Dreams

Wonder Wax

Wonder Wax – basecoat

SURF MORE

XM

X GAMES X LOS ANGELES AUGUST 5-8 2004 • SURFING AT H.B. PIER AUGUST 7 • WWW.EXPN.COM •
X Games
LOS ANGELES
Famous
SURF WAX
FAMOUSWAX.COM
Cool
58° - 68° F · 14.5° - 21° C

X Games

Mr. Zog's Sex Wax – original

Mr. Zog's Sex Wax – quick humps

Zuma Jay – 1970s (left) and 1980s